Alison M. Marr • W.D. Wallis

Magic Graphs

Second edition

Birkhäuser

Alison M. Marr
Department of Mathematics
 and Computer Science
Southwestern University
Georgetown, TX, USA

W.D. Wallis
Department of Mathematics
Southern Illinois University
Carbondale, IL, USA

ISBN 978-0-8176-8390-0 ISBN 978-0-8176-8391-7 (eBook)
DOI 10.1007/978-0-8176-8391-7
Springer New York Heidelberg Dordrecht London

Library of Congress Control Number: 2012950047

Mathematics Subject Classification (2010): 05C78

Preface

Magic Labelings

Magic squares are among the more popular mathematical recreations. Their origins are lost in antiquity; over the years, a number of generalizations have been proposed. In the early 1960s, Sedláček asked whether "magic" ideas could be applied to graphs.

Shortly afterward, Kotzig and Rosa formulated the study of graph labelings or *valuations* as they were first called. A labeling is a mapping whose domain is some set of graph elements—the set of vertices, for example, or the set of all vertices and edges—and whose range was a set of positive integers. Various restrictions can be placed on the mapping. The case that we shall find most interesting is where the domain is the set of all vertices and edges of the graph, and the range consists of the positive integers from 1 up to the number of vertices and edges. No repetitions are allowed.

In particular, one can ask that the set of labels associated with any edge—the label on the edge itself and those on its endpoints—always add to the same sum. Kotzig and Rosa called such a labeling, and the graph possessing it, *magic*. To avoid confusion with the ideas of Sedláček and the many possible variations, we would call it an *edge-magic total labeling*. A related concept, a *vertex-magic total labeling*, is one in which the sum of the label on any vertex and the labels on the edges containing that vertex is always constant. Any labeling that has both these properties (usually with two different constants) is called *totally magic*.

Magic labelings were studied briefly, but they were overshadowed by other graph valuations, in particular by *graceful labelings*, which have a number of applications and are related to the problems of decomposing graphs into trees. However, since the 1990s there has been a resurgence of interest in magic labelings.

One reason for this resurgence is the deceptively simple question, "does every tree have an edge-magic total labeling?" Although it is so easy to ask, no progress has been made toward answering this question for four decades. A number of interesting results about other families of graphs have been discovered, but trees remain elusive.

Several mathematicians have become intrigued by the problem of discovering which graphs are magic. A number of small theorems of combinatorics and graph theory get used in the study of these graphs. For example, in the text, we need to discuss coloring problems and Vizing's Theorem. The construction of magic arrays other than squares is needed. Small structural graph-theoretic theorems need to be invented. And some of the problems seem to be deeper and more difficult than one would at first expect.

Some applications have been studied, mainly in network-related areas. Suppose it is required to assign addresses to the possible links in a communications network. It is required that the addresses are all different, and that the address of a link can be deduced from the identities of the two nodes linked, without the need of using a lookup table. This has been modeled using edge-magic labelings. Another application is in the construction of *ruler models*, which have been applied to the study of *radar pulse codes*.

However, the main reasons for a monograph studying magic labelings are three-fold:

1. Magic labelings provide an introduction to the more general topic of graph labelings. (The topic is large and growing; the interested reader should consult such sources as Gallian's online survey [31] and the book [8] by Bača and Miller on anti-magic labelings.)

2. A focused book, on one particular problem such as this, is a good guide for graduate students beginning research, so they can see how new mathematics comes into existence. In fact, a draft version of the first edition was used as notes for a graduate "special topics" course. Students see some small graph-theoretic proofs and get some idea of how different areas of graph theory interact (as, for example, when Vizing's Theorem on edge-chromatic number is used).

3. In recent years a number of researchers have found the topic fascinating; unfortunately, they have not all communicated very well with each other, and we hope this volume will obviate unnecessary repetition of intellectual effort and help unify the notation, which is currently diverse and self-contradictory.

About this Book

The book begins with a survey of the main ingredients. Magic properties are introduced by a discussion of magic squares, also touching on the related Latin squares and on Latin rectangles, and the basics of graph theory are covered briefly. We then define graph labelings in general and magic labelings in particular. The first chapter also includes a brief sketch of applications. Subsequent chapters explore the three main types of magic labelings—edge-magic, vertex-magic, and totally magic—in turn. Finally, magic labelings of directed graphs are discussed.

Throughout the text there are exercises and research problems. The exercises are designed to aid understanding. Some are quite easy; some ask the reader to do a complete search for labelings of a particular graph or labelings of a particular type; a few are quite difficult. Some of the research problems require very little work, but a few are substantial. A brief commentary on the research problems is included in the volume.

There is an extensive bibliography and solutions to the majority of the exercises. The book closes with an index, in which the convention of italicizing the entries where a definition occurs has been followed.

Some knowledge of groups and fields is assumed in the preliminary chapter, in the discussion of magic squares and Latin squares, but these details can be skipped if desired. Most readers will have a background in graph theory, but a summary has been provided. So there are not many mathematical prerequisites. However, the reader is assumed to have some mathematical maturity, to understand proofs, and to use matrices and modular arithmetic with reasonable facility.

The Second Edition

There have been a number of developments in graph labeling over the last decade—for example, the work on directed graphs—and this is why a new edition seemed appropriate. While every new edition endeavors to contain the most comprehensive and current results, we have chosen to focus on the problems that best introduce readers to magic graphs. Some of the research problems from the first edition have been solved, and we have commented on these results and used them in concocting some new exercises. We are not trying to produce an encyclopedia, so some proofs have been omitted; the reader can always consult the original papers.

Acknowledgments

The study of magic labelings owes a great deal to Alex Rosa and to the late Anton Kotzig, and we are happy to acknowledge their contributions.

Edy Baskoro reminded us of the topic, and David Brown, Jim MacDougall, John McSorley, Mirka Miller, Nick Phillips, Slamin, and the late Gary Bloom joined in extensive discussions and seminars. Thank you all.

Finally, we are grateful for the constant support of Katherine Ghezzi, Tom Grasso, Ann Kostant, and Allen Mann at Birkhäuser.

Contents

List of Figures

1

Preliminaries

1.1 Magic

1.1.1 Magic Squares

Magic squares are among the more popular mathematical recreations. Their origins are lost in antiquity. A classical reference is [2], while one of the better recent books is [12].

A *magic square* of side n is an $n \times n$ array whose entries are an arrangement of the integers $\{1, 2, \ldots, n^2\}$, in which all elements in any row, any column, or either the main diagonal or main back-diagonal, add to the same sum. Examples include

1	15	8	10
12	6	13	3
14	4	11	5
7	9	2	16

1	7	13	19	25
18	24	5	6	12
10	11	17	23	4
22	3	9	15	16
14	20	21	2	8

34	6	5	32	1	33
9	25	11	8	30	28
15	18	20	23	19	16
22	24	14	17	13	21
27	7	26	29	12	10
4	31	35	2	36	3

Variations in the set of entries have frequently been studied—for example, one might ask that the entries all be primes, or all be perfect squares—but we shall

A.M. Marr and W.D. Wallis, *Magic Graphs*, DOI 10.1007/978-0-8176-8391-7_1,
© Springer Science+Business Media New York 2013

only need to discuss cases in which the entries are the first n^2 positive integers. In fact, we usually do not need the constancy of the diagonal and back-diagonal. We shall say a square is *RCmagic* or *semimagic* if all row-sums and column-sums equal the same constant.

1.1.2 Latin Squares

Latin squares are useful in constructing magic squares, and will arise in some other contexts, so we shall briefly survey the main definitions. For more details, see any design theory text, such as [92], or for more details see [23]. A *Latin square* of *order* (or *side*) n is an $n \times n$ array based on some set S of n symbols, with the property that every row and every column contains every symbol exactly once. In other words, every row and every column is a permutation of S. Since the arithmetical properties of the symbols are not used, the nature of the elements of S is immaterial, but it is usually most convenient to use $\{1, 2, \ldots, n\}$.

As an example, let G be a finite group with n elements g_1, g_2, \ldots, g_n. Define an array $A = (a_{ij})$ by $a_{ij} = k$, where $g_k = g_i g_j$. The elements of column j of A are the $g_i g_j$, where g_j is fixed. Now g_j has an inverse element g_j^{-1} in G. If i and k are any two integers from 1 to n, then $g_i g_j = g_k g_j$ would imply $g_i g_j g_j^{-1} = g_k g_j g_j^{-1}$, whence $g_i = g_k$, so $i = k$. Thus, the elements $g_1 g_j, g_2 g_j, \ldots, g_n g_j$ are all different. So column j contains a permutation of $\{1, 2, \ldots, n\}$. A similar proof applies to the rows. So A is a Latin square. This example gives rise to infinitely many Latin squares, including one of the every positive integer order. For example, at orders 1, 2, and 3 we have

$$\boxed{1} \qquad \begin{array}{cc} 1 & 2 \\ 2 & 1 \end{array} \qquad \begin{array}{ccc} 1 & 2 & 3 \\ 2 & 3 & 1 \\ 3 & 1 & 2 \end{array}$$

However, the majority of Latin squares do not come from groups.

Any completed Sudoku puzzle is an example of a 9×9 Latin square. Also, note that although the rows and columns of a completed Sudoku all add to the same value, they are not semimagic squares. In a completed Sudoku puzzle the numbers 1–9 are repeated multiple times (once in each row, column, and subsquare in particular), but in a semimagic square the numbers 1–81 would need to be used without repeats.

Two Latin squares A and B of the same side n are called *orthogonal* if the n^2 ordered pairs (a_{ij}, b_{ij})—the pairs formed by superimposing one square on the other—are all different. We say "A is orthogonal to B" or "B is orthogonal to A"—clearly, the relation of orthogonality is symmetric. More generally, one can speak of a set of k *mutually orthogonal* Latin squares: squares A_1, A_2, \ldots, A_k such that A_i is orthogonal to A_j whenever $i \neq j$.

A set of n cells in a Latin square is called a *transversal* if it contains one cell from each row and one cell from each column and if the n entries include every treatment precisely once. Using this idea we get an alternative definition of orthogonality: A and B are orthogonal if and only if for every treatment x in A, the n cells where A has entry x form a transversal in B. The symmetry of the concept is not so obvious when this definition is used.

If n is a prime power, there exists a set of $n - 1$ mutually orthogonal Latin squares of side n. The standard construction can be described as follows. Suppose the elements of the n-element field are $x_1, x_2, \ldots, x_{n-1} = -1, x_n = 0$. The squares are $A_1, A_2, \ldots, A_{n-1}$, and A_k has (i, j) entry m where $x_m = x_i + x_k x_j$.

There can never be more than $n - 1$ squares of side n in a mutually orthogonal set (except in the trivial case $n = 1$). It is unknown whether this upper bound can ever be achieved when n is not a prime power, but we know there exists a pair of orthogonal squares of every order other than 1, 2, and 6, and a set of three for all orders other than 1, 2, 3, 6, and possibly 10.

Orthogonal Latin squares have been used in the construction of magic squares. If A and B are orthogonal Latin squares of side n based on $\{1, 2, \ldots, n\}$, the array with (i, j) entry $(a_{ij} + n(b_{ij} - 1))$ is semimagic. If $A, B,$ and C are mutually orthogonal Latin squares and C has all diagonal entries equal and all back-diagonal entries equal, the array formed from A and B is magic. The magic square of order 4, given in Sect. 1.1.1, was constructed in this way from the following A, B, C:

1 3 4 2		1 4 2 3		1 2 3 4
4 2 1 3		3 2 4 1		2 1 4 3
2 4 3 1		4 1 3 2		3 4 1 2
3 1 2 4		2 3 1 4		4 3 2 1

For odd orders, the square C cannot have both its main and back diagonals constant, because they have a common member. But if $n = 2t - 1$, one can find orthogonal squares A and B, where A has constant main diagonal (t, t, \ldots, t) and B has constant back diagonal (t, t, \ldots, t), and $(a_{ij} + n(b_{ij} - 1))$ is magic.

The magic square of order 5 in Sect. 1.1.1 was constructed from orthogonal Latin squares, but the order 6 example obviously was not; it was taken from [77].

Exercise 1.1 *Construct a pair of orthogonal Latin squares of order 3, and a magic square of order 3.*

Exercise 1.2 *An $n \times n$ array A is called* circulant *if $a_{i+1,j+1} \equiv a_{ij} + 1 \pmod{n}$ in every case. Use a circulant construction to find a pair of orthogonal Latin squares, and a magic square, of every odd order.*

1.1.3 Magic Rectangles

Magic rectangles are a generalization of magic squares. A *magic rectangle* $A = (a_{ij})$ of size $m \times n$ is an $m \times n$ array whose entries are $\{1, 2, \ldots, mn\}$, each appearing once, with all its row sums, and all its column sums, equal. The sum of all entries in the array is $\frac{1}{2}mn(mn + 1)$; it follows that

$$\sum_{i=1}^{m} a_{ij} = \tfrac{1}{2}n(mn + 1), \text{ all } j,$$

$$\sum_{j=1}^{n} a_{ij} = \tfrac{1}{2}m(mn + 1), \text{ all } i,$$

so m and n must either both be even or both be odd. It was shown in [43, 44] that such an array exists whenever m and n have the same parity, except for the impossible cases where exactly one of m and n is 1, and for $m = n = 2$. Simpler constructions appear in [42].

Magic Rectangles of Size $3 \times m$

For our purposes we need the existence of a $3 \times m$ magic rectangle for every odd m, so for completeness we present constructions (from [42]) for this case. The construction uses smaller arrays as building blocks. For positive integers n and i, define

$$C(i) = \begin{array}{|c|c|} \hline i + 1 & m + 1 - i \\ \hline \tfrac{1}{2}(3m + 1) + i & \tfrac{1}{2}(3m + 1) - i \\ \hline 3n - 2i & 2n + 2i \\ \hline \end{array},$$

$$D(i) = \begin{array}{|c|c|} \hline 3n - 2i & 2n + 2i \\ \hline \tfrac{1}{2}(3m + 1) + i & \tfrac{1}{2}(3m + 1) - i \\ \hline i + 1 & m + 1 - i \\ \hline \end{array}.$$

Theorem 1.1 *There exists a $3 \times m$ magic rectangle whenever m is odd.*

Proof. First suppose $m \equiv 1 \pmod 4$. Define

$$A = \begin{array}{|c|c|c|} \hline 1 & 2n & \tfrac{1}{2}(n + 3) \\ \hline 3n & \tfrac{1}{2}(n + 1) & n + 1 \\ \hline \tfrac{1}{2}(3n + 1) & 2n + 1 & 3n - 1 \\ \hline \end{array}.$$

Then a $3 \times m$ magic rectangle is formed by adjoining to A the $\frac{1}{4}(n - 5)$ arrays $C(i) : 1 \leq i \leq \frac{1}{4}(n-5)$ and the $\frac{1}{4}(n-1)$ arrays $D(j) : \frac{1}{4}(n-1) \leq j \leq \frac{1}{2}(n-3)$.

When $m \equiv 3 \pmod 4$, the case $m = 3$ is easy to construct. So assume $m > 3$, and define t to be $\lfloor \frac{1}{3}(n + 1) \rfloor$. Then $n - 3m = 1, 0$ or -1. Define

$$A = \begin{array}{|c|c|} \hline 1 & \frac{1}{2}(n+3) \\ \hline 3n & n+1 \\ \hline \frac{1}{2}(3n+1) & 3n-1 \\ \hline \end{array},$$

and define an array B by

$$B = \begin{array}{|c|c|c|} \hline m+1 & 2n+1 & 2n+2m \\ \hline \frac{1}{2}(3n+1)+m & \frac{1}{2}(n+1) & \frac{1}{2}(3n+1)-m \\ \hline 3n-2m & 2n & n+1-m \\ \hline \end{array} \text{ when } n-3m=1,$$

$$B = \begin{array}{|c|c|c|} \hline m+1 & 2n & 2n+2m \\ \hline \frac{1}{2}(3n+1)+m & \frac{1}{2}(n+1) & \frac{1}{2}(3n+1)-m \\ \hline 3n-2m & 2n+1 & n+1-m \\ \hline \end{array} \text{ when } n-3m=0,$$

$$B = \begin{array}{|c|c|c|} \hline 3n-2m & 2n+1 & n+1-m \\ \hline \frac{1}{2}(3n+1)+m & \frac{1}{2}(n+1) & \frac{1}{2}(3n+1)-m \\ \hline m+1 & 2n & 2n+2m \\ \hline \end{array} \text{ when } n-3m=-1.$$

The required magic rectangle is formed by adjoining to $A, B,$ and $D(1)$ the $\frac{1}{4}(n-7)$ arrays $C(i), 2 \le i \le \frac{1}{4}(n-3)$ and the $\frac{1}{4}(n-7)$ arrays $D(j), \frac{1}{4}(n+1) \le j \le \frac{1}{2}(n-3), j \ne m$. $\qquad\square$

Exercise 1.3 *Show that the rectangles constructed in Theorem 1.1 are magic.*

Kotzig Arrays

We define a *Kotzig array of size* $m \times n$ to be an $m \times n$ matrix, each row of which is a permutation of $\{0, 1, 2, \ldots, n-1\}$, in which every column has the same sum. The common sum must be $\frac{m}{n}\binom{n}{2} = \frac{1}{2}m(n-1)$, which is an integer only if m is even or n is odd.

An example of a $2 \times n$ Kotzig array is

$$\begin{bmatrix} 0 & 1 & \ldots & n-2 & n-1 \\ n-1 & n-2 & \ldots & 2 & 1 \end{bmatrix}. \tag{1.1}$$

A $3 \times n$ example can exist only if n is odd. If $n = 2r + 1$, the following example is due to Kotzig [52] (which is why we chose the name):

$$\begin{bmatrix} 0 & 1 & \ldots & r & r+1 & r+2 & \ldots & 2r \\ 2r & 2r-2 & \ldots & 0 & 2r-1 & 2r-3 & \ldots & 1 \\ r & r+1 & \ldots & 2r & 0 & 1 & \ldots & r-1 \end{bmatrix}. \tag{1.2}$$

Arrays with more rows can be constructed by appending copies (1.1) and (1.2). So

Lemma 1.2 *There is a Kotzig array of size $m \times n$ whenever $m > 1$ and $m(n-1)$ is even.*

One application is another construction for some magic rectangles. For example, we have

Theorem 1.3 *There exists a $3 \times m$ magic rectangle whenever m is odd and $m \equiv 0 \bmod 3$.*

Proof. Write $m = 3n = 6r + 3$, and denote by A the $3 \times n$ array (1.2). For $j = 1, 2, \ldots, s$, columns $3j - 2, 3j - 1$ and $3j$ of the magic rectangle are

$9j + 1$	$9j + 6$	$9j + 8$
$9(2r - 2j) + 9$	$9(2r - 2j) + 2$	$9(2r - 2j) + 4$
$9(r + j) + 5$	$9(r + j) + 7$	$9(r + j) + 3$

(reduced modulo $3m$ when necessary). In other words, the entries are derived from a 3×3 semimagic square by adding 9 times the element from the relevant row in column j of A. □

Exercise 1.4 *What orders of magic rectangles can be derived from Kotzig arrays, using the ideas of Theorem 1.3?*

1.2 Graphs

The basic ideas of graph theory will be surveyed here, primarily to ensure that writer and readers use all the terminology in the same way. For further details, proofs, etc., the reader should consult a book on the subject (recent examples include [93, 98]).

A *graph* G consists of a finite set $V(G)$ of objects called *vertices* together with a set $E(G)$ of unordered pairs of vertices; the elements of $E(G)$ are called *edges*. We write $v = v(G)$ and $e = e(G)$ for the orders of $V(G)$ and $E(G)$, respectively; these are called the *order* and *size* of G. In terms of the more general definitions sometimes used, we can say that "our graphs are finite and contain neither loops nor multiple edges." A *multigraph* is defined in the same way as a graph except that there may be more than one edge corresponding to the same unordered pair of vertices. Unless otherwise mentioned, all definitions pertaining to graphs will be applied to multigraphs in the obvious way.

The edge containing x and y is written xy or (x, y); x and y are called its *endpoints*. We say this edge *joins* x to y. $G - xy$ denotes the result of deleting

edge xy from G; if x and y were not adjacent then $G+xy$ is the graph constructed from G by adjoining an edge xy. Similarly $G-x$ is the graph derived from G by deleting one vertex x (and all the edges on which x lies). Similarly, $G-S$ denotes the result of deleting some set S of vertices.

If vertices x and y are endpoints of one edge in a graph or multigraph, then x and y are said to be *adjacent* to each other, and it is often convenient to write $x \sim y$. The set of all vertices adjacent to x is called the *neighborhood* of x, and denoted $N(x)$. We define the *degree* or *valency* $d(x)$ of the vertex x to be the number of edges that have x as an endpoint. If $d(x) = 0$, x is an *isolated* vertex, and a vertex x with $d(x) = 1$ is a *leaf* or *pendant vertex*. A graph is called *regular* if all its vertices have the same degree; in particular, if the common degree is 3, the graph is called *cubic*. We write $\delta(G)$ for the smallest of all degrees of vertices of G, and $\Delta(G)$ for the largest. (One also writes $\Delta(G)$ for the common degree of a regular graph G.) If G has v vertices, so that its vertex-set is, say,

$$V(G) = \{x_1, x_2, \ldots, x_v\},$$

then its *adjacency matrix* M_G is the $v \times v$ matrix with entries m_{ij}, such that

$$m_{ij} = \begin{cases} 1 & \text{if } x_i \sim x_j, \\ 0 & \text{otherwise.} \end{cases}$$

A vertex and an edge are called *incident* if the vertex is an endpoint of the edge, and two edges are called incident if they have a common endpoint. A set of edges is called *independent* if no two of its members are incident, while a set of vertices is independent if no two of its members are adjacent.

Theorem 1.4 *In any graph or multigraph, the number of edges equals half the sum of the degrees of the vertices.*

Corollary 1.4.1 *In any graph or multigraph, the number of vertices of odd degree is even. In particular, a regular graph of odd degree has an even number of vertices.*

Given a set S of v vertices, the graph formed by joining all pairs of members of S is called the *complete* graph on S, and denoted K_S. We also write K_v to mean any complete graph with v vertices. The set of all edges of $K_{V(G)}$ that are *not* in a graph G will form a graph with $V(G)$ as vertex-set; this new graph is called the *complement* of G, and written \overline{G}. More generally, if G is a subgraph of H, then the graph formed by deleting all edges of G from H is called the *complement of G in H*, denoted $H-G$. The complement \overline{K}_S of the complete graph K_S on vertex-set S is called a null graph; we also write \overline{K}_v for a null graph with v vertices.

An *isomorphism* of a graph G onto a graph H is a one-to-one map ϕ from $V(G)$ onto $V(H)$ with the property that a and b are adjacent vertices in G if and

only if $\phi(a)$ and $\phi(b)$ are adjacent vertices in H; G is isomorphic to H if and only if there is an isomorphism of G onto H. From this definition it follows that all complete graphs on n vertices are isomorphic. The notation K_n can be interpreted as being a generic name for the typical representative of the isomorphism-class of all n-vertex complete graphs.

If G is a graph, it is possible to choose some of the vertices and some of the edges of G in such a way that these vertices and edges again form a graph, H say. H is then called a *subgraph* of G; one writes $H \le G$. Clearly every graph G has itself and the 1-vertex graph (which we shall denote K_1) as subgraphs; we say H is a *proper* subgraph of G if it neither equals G nor K_1. If U is any set of vertices of G, then the subgraph consisting of U and all the edges of G that joined two vertices of U is called an *induced* subgraph, the *subgraph induced by* U, and is denoted $\langle U \rangle$. A subgraph G of a graph H is called a *spanning* subgraph if $V(G) = V(H)$.

A graph is called *disconnected* if its vertex-set can be partitioned into two subsets, V_1 and V_2, which have no common element, in such a way that there is no edge with one endpoint in V_1 and the other in V_2; if a graph is not disconnected then it is *connected*. A disconnected graph consists of a number of disjoint subgraphs; a maximal connected subgraph is called a *component*.

Partitionings of sets of graph vertices are usually called *colorings*. Instead of talking about a partition of the vertex-set V into three parts V_1, V_2, V_3, we take a set C of three *colors* $\{c_1, c_2, c_3\}$ and define a map $\xi : V \to \{c_1, c_2, c_3\}$. The two approaches are equivalent if V_i is defined as $\{x \mid \xi(x) = c_i\}$.

Formally, suppose $C = \{c_1, c_2, \ldots\}$ is a set of undefined objects called *colors*. A *C-coloring* (or C-vertex coloring) ξ of a graph G is a map

$$\xi : V(G) \to C.$$

The sets $V_i = \{x : \xi(x) = c_i\}$ are called *color classes*. A *proper coloring* of G is a coloring in which no two adjacent vertices belong to the same color class. In other words,

$$x \sim y \Rightarrow \xi(x) \ne \xi(y).$$

A proper coloring is called an *n-coloring* if C has n elements. If G has an n-coloring, then G is called *n-colorable*; 2-colorable and 3-colorable graphs are called *bipartite* and *tripartite* respectively. Tripartite graphs are also called *trichromatic*.

The *complete bipartite graph* on V_1 and V_2 has two disjoint sets of vertices, V_1 and V_2; two vertices are adjacent if and only if they lie in different sets. We write $K_{m,n}$ to mean a complete bipartite graph with m vertices in one set and n in the other. $K_{1,n}$ in particular is called an *n-star*; the vertex of degree n is called the *center*. So a (general) bipartite graph is a subgraph of a complete bipartite graph.

More generally, the *complete r-partite graph* K_{n_1,n_2,\dots,n_r} is a graph with vertex-set $V_1 \cup V_2 \cup \dots \cup V_r$, where the V_i are disjoint sets and V_i has order n_i, in which xy is an edge if and only if x and y are in different sets. Any subgraph of this graph is called an *r-partite* graph. If $n_1 = n_2 = \dots = n_r = n$ we use the abbreviation $K_n^{(r)}$.

A *walk* in a graph G is a finite sequence of vertices x_0, x_1, \dots, x_n and edges a_1, a_2, \dots, a_n of G:

$$x_0, a_1, x_1, a_2, \dots, a_n, x_n,$$

where the endpoints of a_i are x_{i-1} and x_i for each i. A *simple walk* is a walk in which no edge is repeated. A *path* is a walk in which no vertex is repeated; the *length* of a path is its number of edges. A walk is *closed* when the first and last vertices, x_0 and x_n, are equal. A *cycle* of length n is a closed simple walk of length n, $n \geq 3$, in which the vertices x_0, x_1, \dots, x_{n-1} are all different.

A graph that contains no cycles at all is called *acyclic*; a connected acyclic graph is called a *tree*. Clearly all trees are bipartite graphs. Vertices of a tree other than leaves are called *internal*. A union of disjoint trees is called a *forest*.

It is clear that the set of vertices and edges that constitute a path in a graph is itself a graph. We define a walk P_n to be a graph with n vertices x_1, x_2, \dots, x_n and $n-1$ edges $x_1 x_2, x_2 x_3, \dots, x_{n-1} x_n$. A cycle C_n is defined similarly, except that the edge $x_n x_1$ is also included, and (to avoid the triviality of allowing K_2 to be defined as a cycle) n must be at least 3. The latter convention ensures that every C_n has n edges.

A cycle that passes through every vertex in a graph is called a *Hamilton cycle* and a graph with such a cycle is called *Hamiltonian*.

If G is any graph, then a *factor* or *spanning subgraph* of G is a subgraph with vertex-set $V(G)$. A *factorization* of G is a set of factors of G that are pairwise *edge-disjoint*—no two have a common edge—and whose union is all of G.

Every graph has a factorization, quite trivially: since G is a factor of itself, $\{G\}$ is a factorization of G. However, it is more interesting to consider factorizations in which the factors satisfy certain conditions. In particular a *one-factor* is a factor that is a regular graph of degree 1. In other words, a one-factor is a set of pairwise disjoint edges of G that between them contain every vertex. A *one-factorization* of G is a decomposition of the edge-set of G into edge-disjoint one factors. Similarly a *two-factor* is a factor that is a regular graph of degree 2—a union of disjoint cycles —and a *two-factorization* of G is a decomposition of the edge-set of G into edge-disjoint two-factors.

1.3 Digraphs

In many applications, it is useful to associate a direction with each edge.

We shall formally define a *digraph* D to consist of a finite set $V(D)$ of objects called *vertices* together with a set $A(D)$ of ordered pairs of vertices; the elements of $A(D)$ are called *arcs* (or sometimes *directed edges*). An arc directed from vertex x to vertex y is denoted (x, y), and we indicate that such an arc exists by writing $x \rightarrow y$.

The word "digraph" comes from the phrase "directed graph," but observe that a digraph may contain two arcs joining the same pair of vertices, in opposite directions, so the underlying structure, if directions are ignored, may be a multigraph: arcs (x, y) and (y, x) would both correspond to edge xy.

With each vertex x of a digraph we associate two degrees, the *outdegree* $od(x)$, the number of arcs of the form (x, y), and the *indegree* $id(x)$, the number of arcs (z, x). Various terms associated with (undirected) graphs are applied to digraphs in the obvious way; for example, a subdigraph of D is a digraph formed by using some of the vertices and arcs of D. The complete digraph DK_n has n vertices and all $n(n - 1)$ possible arcs. The *double cycle* DC_n is constructed from a cycle of length n by replacing each edge xy by a pair of arcs (x, y) and (y, x).

Further discussion of digraphs will be found in Chap. 5.

1.4 Labelings

A *labeling* (or *valuation*) of a graph is a map that carries graph elements to numbers (usually to the positive or nonnegative integers). The most common choices of domain are the set of all vertices and edges (such labelings are called *total* labelings), the vertex-set alone (*vertex*-labelings), or the edge-set alone (*edge*-labelings). Other domains are possible.

In many cases, it is interesting to consider the sum of all labels associated with a graph label. This will be called the *weight* of the element. For example, the weight of vertex x under labeling λ is

$$wt(x) = \lambda(x) + \sum_{y \sim x} \lambda(xy),$$

while

$$wt(xy) = \lambda(x) + \lambda(xy) + \lambda(y).$$

If necessary, the labeling can be specified by a subscript, as in $wt_\lambda(x)$.

It is sometimes convenient to represent a graph labeling by a *labeling array*. Given a v-vertex graph G with vertices x_1, x_2, \ldots, x_v possessing a labeling λ, the labeling array $L(G)$ is a $v \times v$ incomplete array whose (i, i) entry is $\lambda(x_i)$ and whose (i, j) entry is $\lambda(x_i x_j)$. If x_i is not adjacent to x_j, then the (i, j) cell is empty. If zero labels are not allowed (as will be true in magic labelings) the blanks are replaced by zeroes, and the term "labeling matrix" is also used. Labeling matrices of graphs are of course symmetric, but this will not be so when we discuss digraphs.

We shall define two labelings of the same graph to be *equivalent* if one can be transformed into the other by an automorphism of the graph.

The most complete recent survey of graph labelings is [31].

1.5 Magic Labeling

Various authors have introduced labelings that generalize the idea of a magic square. Sedláček [83] defined a graph to be magic if it had an edge-labeling, with range the real numbers, such that the sum of the labels around any vertex equals some constant, independent of the choice of vertex. These labelings have been studied by Stewart (see, for example, [86]), who called a labeling *supermagic* if the labels are consecutive positive integers. (It is easy to see that for a regular graph, one can assume that this set of integers starts from 1.) Several others have studied these labelings; a recent reference is [32]. Some writers simply use the name "magic" instead of "supermagic" (see, for example, [45]).

Kotzig and Rosa [54] defined a magic labeling to be a total labeling in which the labels are the integers from 1 to $|V(G)| + |E(G)|$ and the sum of labels on an edge and its two endpoints are constant; this constant is called the *magic sum* of the labeling. In 1996 Ringel and Llado [82] redefined this type of labeling (and called the labelings *edge-magic*, causing some confusion with papers that have followed the terminology of [56], mentioned below); see also [33]. Total labelings have also been studied in which the sum of the labels of all edges adjacent to the vertex x, plus the label of x itself, is constant. The first paper on these labelings was [61].

In 1983, Lih [57] introduced magic labelings of planar graphs where labels extended to faces as well as edges and vertices, an idea which he traced back to thirteenth century Chinese roots. Bača (see, for example, [6, 7]) has written extensively on these labelings. A somewhat related sort of magic labeling was defined by Dickson and Rogers in [24].

Lee et al. [56] introduced a weaker concept, which they called *edge-magic*, in 1992. The edges are labeled and the sums at the vertices are required to be congruent modulo the number of vertices.

In order to clarify the terminological confusion defined above, we define a labeling of a graph to be *edge-magic* if the sum of all labels associated with an edge equals a constant independent of the choice of edge, and *vertex-magic* if the same property holds for vertices. (This terminology could be extended to other substructures: face-magic, for example.) The domain of the labeling is specified by a modifier on the word "labeling." We shall always require that the labeling is an injection (one-to-one map) *onto* the appropriate set of consecutive integers starting from 1. For example, Stewart studies vertex-magic edge-labelings (when the graph is regular), and Kotzig and Rosa define edge-magic total labelings.

Any graph with an edge-magic total labeling will be called *edge-magic*. (Some authors refer to a graph as being "edge-magic total." This is not only grammatically painful—the adjective "total" should refer to the mapping, not the graph—but also unnecessary.) Similarly, a graph with a vertex-magic total labeling is called *vertex-magic*.

More recently Enomoto et al. [25] introduced the name *super edge-magic* for magic labelings in the sense of Kotzig and Rosa, with the added property that the v vertices receive the smaller labels, $\{1, 2, \ldots, v\}$. To avoid confusion with the earlier use of "super," some authors call such labelings *strong*, but the usage of "super" seems to have survived. (On another note, Hegde defines strong magic labelings to be edge-magic labelings in which $\underline{k}(G) = \overline{k}(G)$ where $\underline{k}(G)$ is the smallest possible magic sum for a graph G and $\overline{k}(G)$ is the largest [46].) Further confusion arises because "super" has also been applied to labelings in which the edges receive the smaller labels, from 1 to e. In a (probably vain) attempt to clarify the situation, we shall distinguish these two types of edge-magic labelings by calling them V-super and E-super, respectively; when no confusion arises, a super edge-magic labeling will mean a V-super edge-magic labeling. This will be discussed further in Sect. 2.9. Similar language can be used for super vertex-magic labelings and can be found in Sect. 3.10.

Several authors have continued to use the adjective "total" when referring to super-magic labeling. We have even seen the phrase: "the following graphs are super vertex-magic total." This is quite unnecessary: unless both vertices and edges are being labeled, it is meaningless to ask which set of objects receives the smaller labels. So only total labelings can be super-magic.

Our main interest is in edge-magic total labelings, which we sometimes abbreviate to EMTLs, and vertex-magic total labelings (VMTLs). We shall use the term *magic injection* (edge-magic injection, and so on) to denote a labeling with the magic property in which the labels are required to be positive integers but no upper bound is required (and therefore the map is not necessarily onto). If m is the largest label used in a magic injection, and the graph has v vertices and e edges, the difference $m - (v + e)$ is called the *deficiency* of the injection.

It is conceivable that the same labeling could be both vertex-magic and edge-magic for a given graph (not necessarily with the same constant).

In that case the labeling, and the graph, are called *totally magic*. Totally magic graphs appear to be very rare. For this reason, totally magic injections have also been studied, and graphs with a totally magic injection will be referred to as TMI graphs.

There are a number of interesting types of magic labelings of digraphs. Note that some of these labelings will not be total labelings as studying edge and vertex labelings of digraphs turns out to be interesting and useful in the study of total labelings as well. We shall discuss them in Chap. 5. As far as possible we shall stay close to the above definitions.

1.6 Some Applications of Magic Labelings

1.6.1 Efficient Addressing Systems

The problem of efficient addressing systems is introduced in [15]. Suppose it is necessary to assign addresses to the possible links in a communications network. It is required that the addresses are all different, and that the address of a link can be deduced from the identities of the two nodes linked, without the need of using a lookup table.

The solution proposed in [15] is as follows. First, a graph (the *underlying graph* of the network) is constructed with the nodes as vertices and edges between all pairs of nodes where a link is provided. The graph vertices are labeled in such a way that the differences between endpoints of edges are all distinct—this is called a *semi-graceful* labeling. Then the address of a link is the difference between the labels on its endpoints.

The following solution, using an edge-magic total labeling, provides some additional information. Suppose the underlying graph has an edge-magic total labeling λ with edge-weight k. The nodes and links are assigned the labels of the corresponding vertices and edges. Then the address of the link from x to y is readily calculated as $k - \lambda(x) - \lambda(y)$. Moreover a unique address is available for messages from the system operator to the nodes: the link from the sysop to node x receives address $\lambda(x)$.

If no edge-magic total labeling of the underlying graph is available, one can use an edge-magic injection. An injection with small excess will lead to a smaller range of possible addresses.

1.6.2 Ruler Models

Suppose λ is an edge-magic injection of K_n. It is easy to see that, for any vertices x, y, z, t of K_n, $\lambda(xy) = k - \lambda(x) - \lambda(y)$ and $\lambda(zt) = k - \lambda(z) - \lambda(t)$, so

$\lambda(x)+\lambda(y)$ cannot equal $\lambda(z)+\lambda(t)$. Thus, $\lambda(x)-\lambda(z) \neq \lambda(t)-\lambda(y)$. Therefore the $\binom{n}{2}$ differences between the labels of K_n are all different.

If λ is any labeling of K_n, a *ruler model* of λ is constructed as follows. For each vertex of K_n, place a mark distance $\lambda(x)$ from the start of the ruler. The ruler can be used to measure all distances corresponding to the distance between two marks. Ruler models are discussed, for example, in [14, 15].

The ruler models derived from edge-magic injections have the following special property. No two of the measurable differences are equal, except possibly for differences of the form $\lambda(x) - \lambda(y)$ and $\lambda(y) - \lambda(z)$. This is very nearly the semi-graceful property discussed in the preceding section, so magic sublabelings can be used in many of the situations where semi-graceful labelings have been applied.

One important application of semi-graceful labelings is to *radar pulse codes*. Radar distance ranging is accomplished by transmitting a pulse or train of pulses and waiting for its return after reflection. Only a small fraction of the transmitted energy ever returns to the detector. Because accuracy in measuring target distance is determined by accuracy in measuring the time until the reflected signal is received, it is desirable to have a very narrow transmitted radar pulse whose moment of return can be measured precisely. High-energy pulses are broader than low-energy pulses. On the other hand, a low-energy pulse may be too weak to detect after reflection. The solution is to send a set of low-amplitude, narrowly defined pulses, and to increase the total energy when necessary by increasing the number of pulses.

The most efficient way to proceed is to time the pulses in accordance with the marks on a ruler derived from a semi-graceful labeling. An array of detectors is distributed like a template of the transmitted pulse-train. The signal will match up precisely with the detectors when it returns. At no other time more than one signal will excite a detector. If the ruler from a magic sublabeling is used, there may be two detectors excited at once (if $\lambda(x) - \lambda(y) = \lambda(y) - \lambda(z)$ occurs) but it is still very unlikely that the reflected pulse will be misidentified, even if part of the signal is lost through dispersion.

In this application, the duration of the train will be minimized if the size of the largest label is minimized, so again sublabelings with small excess are better.

2

Edge-Magic Total Labelings

2.1 Basic Ideas

2.1.1 Definitions

An *edge-magic total labeling* on a graph G is a one-to-one map λ from $V(G) \cup E(G)$ onto the integers $1, 2, \ldots, v + e$, where $v = |V(G)|$ and $e = |E(G)|$, with the property that, given any edge xy,

$$\lambda(x) + \lambda(xy) + \lambda(y) = k$$

for some constant k. In other words, $\mathrm{wt}(xy) = k$ for any choice of edge xy. Then k is called the *magic sum* of G. Any graph with an edge-magic total labeling will be called *edge-magic*.

As an example of edge-magic total labelings, Fig. 2.1 shows an edge-magic total labeling of $K_4 - e$.

An edge-magic total labeling will be called *(V-)super edge-magic* if it has the property that the vertex-labels are the integers $1, 2, \ldots, v$, the *smallest* possible labels. A graph with a super edge-magic total labeling will be called *super edge-magic*. Recall that some authors may call such a edge-magic labeling *strong* and the resulting graph a *strongly edge-magic graph*.

A.M. Marr and W.D. Wallis, *Magic Graphs*, DOI 10.1007/978-0-8176-8391-7_2,
© Springer Science+Business Media New York 2013

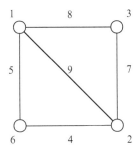

Fig. 2.1. An edge-magic total labeling of $K_4 - e$ with $k = 12$

2.1.2 Some Elementary Counting

As a standard notation, assume the graph G has v vertices $\{x_1, x_2, \ldots, x_v\}$ and e edges. For convenience, we always say vertex x_i has degree d_i and receives label a_i. As we shall frequently refer to the sum of consecutive integers, we define

$$\sigma_i^j = (i+1) + (i+2) + \cdots j = i(j-i) + \binom{j-i+1}{2}. \tag{2.1}$$

The necessary conditions in order that $\{a_1, a_2, \ldots, a_v\} = \lambda(V(G))$, where λ is an edge-magic total labeling of a graph G with magic sum k, are

1. $a_h + a_i + a_j = k$ cannot occur if any two of x_h, x_i, x_j are adjacent.

2. The sums $a_i + a_j$, where $x_i x_j$ is an edge, are all distinct.

3. $0 < k - (a_i + a_j) \leq v + e$ when x_i is adjacent to x_j.

Suppose λ is an edge-magic labeling of a given graph with sum k. If x and y are adjacent vertices, then edge xy has label $k - \lambda(x) - \lambda(y)$. Since the sum of all these labels plus the sum of all the vertex labels must equal the sum of the first $v + e$ positive integers, k is determined. So the vertex labels specify the complete labeling.

Of course, not every possible assignment will result in an edge-magic labeling: the above process may give a non-integral value for k, or give repeated labels.

Among the labels, write S for the set $\{a_i : 1 \leq i \leq v\}$ of vertex labels, and s for the sum of elements of S. Then S can consist of the v smallest labels, the v largest labels, or somewhere in between, so

$$\sigma_0^v \leq s \leq \sigma_e^{v+e},$$

$$\binom{v+1}{2} \leq s \leq ve + \binom{v+1}{2}. \tag{2.2}$$

Clearly, $\sum_{xy \in E}(\lambda(xy) + \lambda(x) + \lambda(y)) = ek$. This sum contains each label once, and each vertex label a_i an additional $d_i - 1$ times. So

$$ke = \sigma_0^{v+e} + \sum (d_i - 1)a_i. \qquad (2.3)$$

If e is even, every d_i is odd and $v + e \equiv 2(\mathrm{mod}\,4)$ then (2.3) is impossible. We have

Theorem 2.1 [82] *If G has e even and $v + e \equiv 2(\mathrm{mod}\,4)$, and every vertex of G has odd degree, then G has no edge-magic total labeling.*

(A generalization of this Theorem will be proven in Sect. 2.2).

Corollary 2.1.1 *The complete graph K_v is not magic when $v \equiv 4(\mathrm{mod}\,8)$. The n-spoke wheel W_n, formed from C_n by adding a new vertex and joining it to every existing vertex, is not magic when $n \equiv 3(\mathrm{mod}\,4)$.*

(We shall see in Sect. 2.3.2 that K_n is never magic for $n > 6$, so the first part of the Corollary really only eliminates K_4.)

If G is super edge-magic, then S must consist of the first v integers. In many cases equation (2.3) can be used to show that this is impossible, because there is no assignment of these integers to the vertices such that $\sigma_0^{v+e} + \sum (d_i - 1)a_i$ is divisible by e. For example, for the cycle C_v where v is even, (2.3) is

$$kv = \sigma_0^{2v} + \sigma_0^v = v(2v + 1) + \tfrac{1}{2}v(v + 1),$$

so $k = 2v + 1 + \tfrac{1}{2}(v + 1)$, which is not integral. Thus no even cycle is super edge-magic.

Exercise 2.1 *Prove that the graph tK_4, consisting of t disjoint copies of K_4, has no edge-magic total labeling when t is odd. Is the same true of tK_v when $v \equiv 4(\mathrm{mod}\,8)$?*

Exercise 2.2 *Prove that the union of t n-wheels, tW_n, has no edge-magic total labeling when t is odd and $n \equiv 3(\mathrm{mod}\,4)$.*

Research Problem 2.1 *Investigate graphs G for which equation (2.3) implies the nonexistence of an edge-magic total labeling of $2G$.*

Equation (2.3) may be used to provide bounds on k. Suppose G has v_j vertices of degree j, for each i up to Δ, the largest degree represented in G. Then ke cannot be smaller than the sum obtained by applying the v_Δ smallest labels to the vertices of degree Δ, the next-smallest values to the vertices of degree $\Delta - 1$, and so on; in other words,

$$ke \geq (d_\Delta - 1)\sigma_0^{v_\Delta} + (d_{\Delta-1} - 1)\sigma_{v_\Delta}^{v_\Delta + (v_{\Delta-1})} + \cdots$$

$$+ \sigma_{v_\Delta + (v_{\Delta-1}) + \cdots + v_3}^{v_\Delta + (v_{\Delta-1}) + \cdots + v_2} + \binom{v + e + 1}{2}.$$

An upper bound is achieved by giving the *largest* labels to the vertices of highest degree, and so on.

Exercise 2.3 *Suppose a regular graph G of degree d is edge-magic. Prove*

$$ke = (d-1)s + \sigma_0^{v+e} = (d-1)s + \tfrac{1}{2}(v+e)(v+e+1), \quad (2.4)$$
$$kdv = 2(d-1)s + (v+e)(v+e+1). \quad (2.5)$$

Avadayappan, Vasuki and Jeyanthi [5] define the *magic strength* of an edge-magic graph G to be the smallest value k such that G has a magic labeling with magic sum k. From equations (2.4) and (2.5) we deduce that for an edge-magic regular graph

$$ke \leq \left(\frac{(d-1)v(v+1) + (v+e)(v+e+1)}{2} \right),$$

so

Theorem 2.2 *An edge-magic regular graph of degree d, with v vertices and e edges, has magic sum at least*

$$\left\lceil \frac{(d-1)v(v+1) + (v+e)(v+e+1)}{2e} \right\rceil. \quad (2.6)$$

Exercise 2.4 *Prove that the magic strength of a cycle C_n is at least $\lceil \tfrac{1}{2}(5v+3) \rceil$.*

2.1.3 Duality

Given a labeling λ, its *dual* labeling λ' is defined by

$$\lambda'(x_i) = (v + e + 1) - \lambda(x_i),$$

and for any edge xy,

$$\lambda'(xy) = (v + e + 1) - \lambda(xy).$$

It is easy to see that if λ is a magic labeling with magic sum k, then λ' is a magic labeling with magic sum $k' = 3(v + e + 1) - k$. The sum of vertex labels is $s' = v(v + e + 1) - s$.

Either s or s' will be less than or equal to $\tfrac{1}{2}v(v + e + 1)$. This means that, in order to see whether a given graph has an edge-magic total labeling, it suffices to check either all cases with $s \leq \tfrac{1}{2}v(v+e+1)$ or all cases with $s \geq \tfrac{1}{2}v(v+e+1)$ (equivalently, check either all cases with $k \leq \tfrac{3}{2}(v + e + 1)$ or all with $k \geq \tfrac{3}{2}(v + e + 1)$).

2.2 Graphs with no Edge-Magic Total Labeling

The results of this section are from the unpublished technical report [51].

2.2.1 Main Theorem

Theorem 2.3 *Suppose G is a graph with v vertices and e edges, where e is even, and suppose every vertex of G has odd degree. Select a positive integer δ such that for each vertex x_i*

$$d(x_i) = 2^\delta d_i + 1$$

for some nonnegative integer d_i. If $T = \sum d_i$, define τ and Q by

$$T = 2^\tau \cdot Q, \ \tau \text{ integral}, \ Q \text{ odd}.$$

If G has an edge-magic total labeling λ, then

$$\begin{aligned}
\tau = 0 &\Rightarrow v = 2^\delta V \text{ for some } V \equiv 1 \bmod 2 \\
&\quad\ e = 2^\delta E \text{ for some } E \equiv 1 \bmod 2 \\
\tau = 1 &\Rightarrow v = 2^{\delta+1} V' \text{ for some } V' \equiv 1 \bmod 2 \\
&\quad\ e = 2^{\delta+1} E' \text{ for some } E' \equiv 1 \bmod 2 \\
\tau \geq 2 &\Rightarrow 2^{\delta+2} \text{ divides } v \text{ and } 2^{\delta+2} \text{ divides } e
\end{aligned}$$

Proof. Since e is even, let us write $e = 2^\nu E$ for some odd integer E. The familiar equation $2e = \sum d(x_i)$ yields

$$2e = v + 2^\delta T,$$

so

$$v + e = 3e - 2^{\delta+\tau} \cdot Q = 2^\nu \cdot 3E - 2^{\delta+\tau} \cdot Q.$$

So (2.3) is

$$(2^\nu \cdot 3E - 2^{\delta+\tau} \cdot Q)(2^\nu \cdot 3E - 2^{\delta+\tau} \cdot Q + 1)/2 = 2^\nu \cdot Ek - 2^\delta \cdot \sum d_i \lambda(x_i),$$

which is of the form

$$2^\delta \cdot (2R - 2^\tau \cdot X) = 2^\nu \cdot Y$$

where $R = \sum d_i \lambda(x_i)$, $X = (2^{\delta+\tau} Q - 3 \cdot 2^{\nu+1} - 1)Q$ and $Y = 2Ek - 2^\nu \cdot 9E^2 - 3E$. The actual values of X and Y are unimportant; what matters is that both are odd, and the result follows. $\qquad\square$

Corollary 2.3.1 *Suppose $e(G)$ is even and $e(G) + v(G) \equiv 2(mod\ 4)$, and suppose each vertex of G has odd degree. Then G is not edge-magic.*

2.2.2 Forests

Theorem 2.4 *Suppose G is a forest with c component trees and suppose G satisfies the conditions of Theorem 2.3. Then*

$$\tau = 0 \Rightarrow c \equiv 0 \bmod 2^{\delta+1} \text{ and } e \equiv 2^{\delta} \bmod 2^{\delta+1}$$
$$\tau = 1 \Rightarrow c \equiv 0 \bmod 2^{\delta+2} \text{ and } e \equiv 2^{\delta+1} \bmod 2^{\delta+2}$$
$$\text{or } c \equiv 2^{\delta+1} \bmod 2^{\delta+2} \text{ and } e \equiv 0 \bmod 2^{\delta+2}$$
$$\tau \geq 2 \Rightarrow c \equiv e \equiv x \bmod 2^{\delta+2} \text{ where } x = 0 \text{ or } 2^{\delta+1}.$$

Proof. For a forest, $v = c + e$; the results follow from Theorem 2.3. □

A family \mathcal{F} of forests is defined as follows. If δ is any positive integer, a δ-tree is a tree in which each vertex x_i has degree $d(x_i) = 1 + 2^{\delta} d_i$ for some nonnegative integer d_i, and a δ-forest is a forest *with an even number of edges* in which every component tree is a δ-tree. Figure 2.2 shows a 2-forest—the vertices have degrees $1, 5 (= 2^2 + 1)$ and $9 (= 2^3 + 1)$.

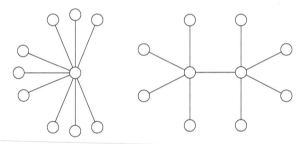

Fig. 2.2. A 2-forest

If F is a δ-forest define $\tau(F) = \tau$ to be the nonnegative integer such that $\sum_i d_i = 2^{\tau} \cdot T$ where T is odd, then \mathcal{F}_{δ} is the set of all δ-forests F that satisfy one of the following conditions, where e is the total number of edges and c is the number of component trees of F:

(i) $\tau(F) = 0$ and $c \not\equiv 0 \bmod 2^{\delta+1}$ or $e \not\equiv 0 \bmod 2^{\delta}$;

(ii) $\tau(F) = 1$ and $c \not\equiv 0 \bmod 2^{\delta+1}$ or $e \not\equiv 0 \bmod 2^{\delta+1}$ or $c + e \not\equiv 2^{\delta+1}$ mod $2^{\delta+2}$;

(iii) $\tau(F) \geq 2$ and $c \not\equiv e \bmod 2^{\delta+2}$ or $c \not\equiv 0 \bmod 2^{\delta+1}$.

Finally, \mathcal{F} is the union of all the sets \mathcal{F}_{δ}.

Theorem 2.5 *For every nonnegative integer τ there are infinitely many forests F belonging to \mathcal{F} that have $\tau(F) = \tau$.*

Proof. Select natural numbers p, q and t such that $p \geq q$. F_{pqt} is the disjoint union of $(2t + 1) \cdot 2^q$ copies of the star with $2^p + 1$ edges. This forest has $c = (2t + 1) \cdot 2^q$ components. Vertex x_i of F_{pqt} has degree $2^p d_i + 1$, where $d_i = 0$ or 1 and $\sum d_i = (2t + 1) \cdot 2^q$. In terms of the definition of \mathcal{F}, $\delta = p$ and $\tau = q$. Clearly c is not divisible by 2^{p+1}, since the largest power of 2 dividing c is 2^q, so $F_{pqt} \in \mathcal{F}$. So for any natural number τ we have constructed infinitely many members of \mathcal{F} with that τ-value.

For $\tau = 0$, consider the graph F_n consisting of the disjoint union of a 5-star with $(4n - 3)$ 3-stars. One vertex has degree $2^1 \cdot 2 + 1$ ($d_i = 2$), $4n - 3$ vertices have degree $2^1 \cdot 1 + 1$ ($d_i = 1$), and the other vertices have degree $2^1 \cdot 0 + 1$ ($d_i = 0$). So $\delta = 1$ and $\tau = 0$. Moreover $c = 4x - 2$ is not divisible by $4 = 2^{\delta+1}$, so $F_x \in \mathcal{F}$, and we have constructed infinitely many members of \mathcal{F} that have $\tau = 0$. □

It is clear from Theorem 2.4 that a member of \mathcal{F} cannot have an edge-magic total labeling. So we have:

Corollary 2.5.1 *There are infinitely many forests with no edge-magic total labeling.*

2.2.3 Regular Graphs

Theorem 2.6 *Suppose G is a regular graph of odd degree d with v vertices and e edges, where e is even. Write*

$$d = 2^r s + 1$$

where s is an odd positive integer and r is a positive integer. If G is edge-magic, then 2^{r+2} divides v.

Proof. As usual, $2e = dv$. Since e is even and d is odd, 4 divides v; write $v = 4V$, so that $e = 2dV$. Then (2.3) becomes

$$(4V + 2dV)(4V + 2dV + 1)/2 = 2dVk - (d - 1)\sum_x \lambda(x),$$

and

$$V[(d + 2)(4V + 2dV + 1) - 2dk] = 2^r s \sum_x \lambda(x).$$

The coefficient of V on the left-hand side is odd, so 2^r divides V, giving the result. □

Remark. This theorem could also be expressed as follows. If a regular graph with v vertices is edge-magic, then

$$d \equiv 3 \,(\mathrm{mod}\ 4) \qquad\qquad \Rightarrow v \equiv 2 \,(\mathrm{mod}\ 4) \text{ or } v \equiv 0 \,(\mathrm{mod}\ 8)$$
$$d \equiv 5 \,(\mathrm{mod}\ 8) \qquad\qquad \Rightarrow v \equiv 2 \,(\mathrm{mod}\ 4) \text{ or } v \equiv 0 \,(\mathrm{mod}\ 16)$$
$$\cdots$$
$$d \equiv 2^{p-1} + 1 \,(\mathrm{mod}\ 2^p) \Rightarrow v \equiv 2 \,(\mathrm{mod}\ 4) \text{ or } v \equiv 0 \,(\mathrm{mod}\ 2^{p+1}).$$

A family \mathcal{R} of regular graphs is now defined as follows. If G is regular of degree d and has v vertices, then $G \in \mathcal{R}$ if and only if there exist natural numbers α, β, γ and δ satisfying

$$\delta < 2^{\alpha},$$
$$d = 2^{\alpha+1}\beta - 2^{\alpha} + 1,$$
$$v = 4(2^{\alpha} \cdot \gamma - \delta).$$

Theorem 2.7 *No member of \mathcal{R} is edge-magic.*

Proof. Suppose α, β, γ, and δ are natural numbers and $\delta < 2^{\alpha}$. Consider a regular graph G of degree $d = 2^{\alpha+1}\beta - 2^{\alpha} + 1$ with $v = 4(2^{\alpha} \cdot \gamma - \delta)$ vertices (a typical element of \mathcal{R}). Writing $r = \alpha$ and $s = 2\beta - 1$ we have $d = 2^r s + 1$ so by Theorem 2.6 G can have an edge-magic total labeling only if v is divisible by $2^{r+2} = 2^{\alpha+2}$. However $v \equiv -4\delta \not\equiv 0 \,(\mathrm{mod}\ 2^{\alpha+2})$ as $0 < \delta < 2^{\alpha}$. So G has no edge-magic total labeling. □

Lemma 2.8 *If d is any odd natural number greater than 1, there are infinitely many graphs of degree d that belong to \mathcal{R}.*

Remark. Lemma 3 of [51] allows the case $d = 1$. The first line of the proof there is "For the case $d = 1$, lemma 4 (*sic*) follows immediately from Theorem 3." (His Theorem 3 is our Theorem 2.6.) However, if $d = 1$, the definition of \mathcal{R} would require $2^{\alpha+1}\beta = 2^{\alpha}$, which is impossible for natural β.

Proof of Lemma 2.8. We construct solutions in the special case $\delta = 1$ (so that $\delta < 2^{\alpha}$ is obviously true).

Assume d is an odd positive integer greater than 1. There exist unique integers α and b such that b is odd and $d = 2^{\alpha} b + 1$. Define $\beta = (b + 1)/2$. Then $d = 2^{\alpha+1}\beta - 2^{\alpha} + 1$.

If d is any natural number and v is any even integer greater than d, there exists a regular graph of degree d on v vertices (select any one-factorization of K_v and delete any $v - d - 1$ factors). If d is odd, then among the even integers greater than d, there will be infinitely many of the form $4(2^{\alpha} \cdot \gamma - 1)$ where α was defined above, and each gives rise to a member of \mathcal{R}. □

Corollary 2.8.1 *If d is any odd natural number, there are infinitely many regular graphs of degree d that have no edge-magic total labeling.*

Proof. For $d = 1$, the graphs eK_2, e even, suffice. For larger values of d, use Lemma 2.8. □

2.3 Cliques and Complete Graphs

2.3.1 Sidon Sequences and Labelings

Suppose the graph G has an edge-magic total labeling λ, and suppose G contains a complete subgraph (or *clique*) H on n vertices. Let us write x_1, x_2, \ldots, x_n for the vertices of H, and denote $\lambda(x_i)$ by a_i. Without loss of generality we can assume the names x_i to have been chosen so that $a_1 < a_2 < \cdots < a_n$.

If k is the magic sum, then $\lambda(x_i x_j) = k - a_i - a_j$, so the sums $a_i + a_j$ must all be distinct. This property is called being *well-spread*; in particular, a *Sidon sequence* or *well-spread sequence* $A = (a_1, a_2, \ldots, a_n)$ of length n is a sequence with the properties:

1. $0 < a_1 < a_2 < \cdots < a_n$.

2. If a_i, a_j, a_k, a_ℓ are all different, then $a_i + a_j \neq a_k + a_\ell$.

Such sequences are related to the work of Sidon [84]. Their study was initiated by Erdös and Turan [27]. A survey of work on them appears in [41].

In discussing Sidon sequences (or, equivalently, cliques in edge-magic graphs), the difference $d_{ij} = |a_j - a_i|$ (the absolute difference between the labels on the endpoints of the edge $x_i x_j$) will be important.

Lemma 2.9 *Suppose A is a Sidon sequence of length n. If $d_{ij} = d_{pq}$, then $\{a_i, a_j\}$ and $\{a_p, a_q\}$ have a common member. No three of the differences d_{ij} are equal.*

Proof. Suppose $d_{ij} = d_{pq}$. We can assume that $i > j$ and $p > q$. If $p = i$, we get $a_i = a_p$ and $a_j = a_q$ and the result is trivial. Thus, without loss of generality we can also assume $p > i$. Then $a_i - a_j = a_p - a_q$, so $a_i + a_q = a_j + a_p$. Therefore $a_i = a_q$ and $i = q$ ($a_j = a_p$ is impossible), and $p > i > j$—the common element is the middle one in order of magnitude.

Now suppose three pairs have the same difference. By the above reasoning there are two possibilities: the pairs must have a common element, or form a triangle. In the former case, suppose the differences are d_{ij}, d_{ik} and $d_{i\ell}$.

From $d_{ij} = d_{ik}$ we must have either $k > i > j$ or $j > i > k$; let's assume the former. Then $d_{ij} = d_{i\ell}$ implies $i > j > \ell$. So j is greater than both k and ℓ. But $d_{ik} = d_{i\ell}$ must mean that either $k > j > \ell$ or $\ell > j > k$, both of which are impossible. On the other hand, suppose they form a triangle, say $d_{ij} = d_{ik} = d_{jk}$. We can assume $i > j$. Then $d_{ij} = d_{jk}$ implies $i > j > k$, and $j > k$ and $d_{ik} = d_{jk}$ imply $k > i$, again a contradiction. $\qquad\square$

Lemma 2.10 *Suppose A is a Sidon sequence of length n. If $d_{ij} = d_{ik}$, then $d_{ij} \leq \frac{1}{2}d_{1n}$.*

Proof. Suppose $d_{ij} = d_{ik}$, and assume $j < k$. Then $d_{jk} = a_k - a_j = a_i - a_j + a_k - a_i = d_{ij} + d_{ik} = 2d_{ij}$. But $a_1 \leq a_j$ and $a_k \leq a_n$, so $d_{jk} \leq d_{1n}$, giving the result. $\qquad\square$

Theorem 2.11 *In any Sidon sequence of length n, $\binom{n}{2} \leq \lfloor\frac{3}{2}d_{1n}\rfloor$, or equivalently $d_{1n} \geq \lceil\frac{1}{3}n(n-1)\rceil$.*

Proof. There are $\binom{n}{2}$ unordered pairs of elements in the sequence, so there are $\binom{n}{2}$ differences. From Lemmas 2.9 and 2.10, the collection of values of these differences can contain the integers $1, 2, \ldots, \lfloor\frac{1}{2}r\rfloor$ at most twice each, and $\lfloor\frac{1}{2}r\rfloor + 1, \ldots, d_{1n}$ at most once each. The result follows. $\qquad\square$

2.3.2 Complete Graphs

Theorem 2.12 [90] *Suppose K_v has an edge-magic total labeling with magic sum k. The number p of vertices that receive even labels satisfies the following conditions:*

(i) *If $v \equiv 0$ or $3 \pmod 4$ and k is even, then $p = \frac{1}{2}(v - 1 \pm \sqrt{v+1})$.*

(ii) *If $v \equiv 1$ or $2 \pmod 4$ and k is even, then $p = \frac{1}{2}(v - 1 \pm \sqrt{v-1})$.*

(iii) *If $v \equiv 0$ or $3 \pmod 4$ and k is odd, then $p = \frac{1}{2}(v + 1 \pm \sqrt{v+1})$.*

(iv) *If $v \equiv 1$ or $2 \pmod 4$ and k is odd, then $p = \frac{1}{2}(v + 1 \pm \sqrt{v+3})$.*

Proof. Suppose λ is an edge-magic total labeling of K_v with magic sum k. Let V_e denote the set of all vertices x such that $\lambda(x)$ is even, and V_o the set of vertices x with $\lambda(x)$ odd; define p to be the number of elements of V_e. Write E_1 for the set of edges with both endpoints in the same set, either V_o or V_e, and E_2 for the set of edges joining the two vertex-sets, so that $|E_1| = \binom{p}{2} + \binom{v-p}{2}$ and $|E_2| = p(v-p)$.

If k is even, then $\lambda(yz)$ is even whenever yz is an edge in E_1 and odd when yz is in E_2, so there are precisely $p + \binom{p}{2} + \binom{v-p}{2}$ even labels. But these labels must

be the even integers from 1 to $\binom{v+1}{2}$, taken once each; so

$$p + \binom{p}{2} + \binom{v-p}{2} = \left\lfloor \frac{1}{2}\binom{v+1}{2} \right\rfloor. \tag{2.7}$$

If $\binom{v+1}{2}$ is even, this equation has solutions $p = \frac{1}{2}(v - 1 \pm \sqrt{v+1})$, while $\binom{v+1}{2}$ odd gives solutions $p = \frac{1}{2}(v - 1 \pm \sqrt{v-1})$.

If k is odd, the edges in E_1 are those that receive the odd labels, and instead of (2.7) we have

$$p + p(v - p) = \left\lfloor \frac{1}{2}\binom{v+1}{2} \right\rfloor \tag{2.8}$$

with solutions $p = \frac{1}{2}(v + 1 \pm \sqrt{v+1})$ when $\binom{v+1}{2}$ is even and $p = \frac{1}{2}(v + 1 \pm \sqrt{v+3})$ when $\binom{v+1}{2}$ is odd.

Using the fact that $\binom{v+1}{2}$ is even when $v = 0$ or $3 \pmod 4$ and odd otherwise, we have the result. □

Now p must be an integer, so the functions whose roots are taken must always be perfect squares. Therefore:

Corollary 2.12.1 *Suppose K_v has an edge-magic total labeling. If $v \equiv 0$ or $3 \pmod 4$, then $v + 1$ is a perfect square. If $v \equiv 1$ or $2 \pmod 4$, then either $v - 1$ is a perfect square and the magic sum of the labeling is even, or $v + 3$ is a perfect square and the magic sum of the labeling is odd.*

This corollary rules out edge-magic total labelings of K_4 (again!—see Corollary 2.1.1) and K_7, as well as infinitely many larger values. The larger values, however, will all be excluded by the next theorem.

Suppose there is an edge-magic total labeling of K_v, where $v \geq 8$. The vertex labels will form a Sidon sequence of length v, A say. Let us denote the edge labels b_1, b_2, \ldots, b_e, where $b_1 < b_2 < \cdots < b_e$; of course, $e = \binom{v}{2}$. If the magic sum is k, then

$$k = a_1 + a_2 + b_e \tag{2.9}$$
$$= a_1 + a_3 + b_{e-1} \tag{2.10}$$
$$= a_v + a_{v-1} + b_1 \tag{2.11}$$
$$= a_v + a_{v-2} + b_2. \tag{2.12}$$

Subtracting (2.9) from (2.10),

$$a_3 - a_2 = b_e - b_{e-1}, \tag{2.13}$$

while (2.11) and (2.12) yield

$$a_{v-1} - a_{v-2} = b_2 - b_1. \tag{2.14}$$

Suppose labels $1, 2, v+e-1$ and $v+e$ were all edge labels. Then $b_1 = 1, b_2 = 2, b_{e-1} = v+e-1$ and $b_e = v+e$. So, from (2.13) and (2.14), $a_3 - a_2 = a_{v-1} - a_{v-2} = 1$. But $2, 3, v-2$ and $v-1$ are all distinct, so this contradicts Lemma 2.9. So one of $1, 2, v+e-1, v+e$ is a vertex label. Without loss of generality we can assume either 1 or 2 is a vertex label (otherwise, the dual labeling will have this property). So $a_1 = 1$ or 2.

Equations (2.9) and (2.11) give

$$a_v = b_e - (a_{v-1} - a_2) - (b_1 - a_1). \tag{2.15}$$

Since $(a_2, a_3, \ldots, a_{v-1})$ is a Sidon sequence of length $v-2$ (any subsequence of a Sidon sequence is also well-spread), Lemma 2.11 applies to it, and $(a_{v-1} - a_2) \geq \lceil \frac{1}{3}(v-2)(v-3) \rceil$, which is at least 10 because $v \geq 8$. Also $(b_1 - a_1) \geq -1$ ($b-1$ is at least 1 and a_1 is at most 2), and $b_e \leq v+e$. So, from (2.15),

$$a_v \leq v + e - 9.$$

So the six largest labels are all edge labels:

$$b_{e-5} = v+e-5, b_{e-4} = v+e-4, \ldots, b_e = v+e.$$

From (2.9) and (2.10), we get

$$k = a_1 + a_2 + v + e = a_1 + a_3 + v + e - 1,$$

so $a_3 = a_2 + 1$. The next smallest sum of two vertex-labels, after $a_1 + a_2$ and $a_1 + a_3$, may be either $a_2 + a_3$ or $a_1 + a_4$.

If it is $a_2 + a_3$, then

$$k = a_2 + a_3 + v + e - 2$$

and by comparison with (2.10), $a_2 = a_1 + 1$. The next-smallest sum is $a_1 + a_4$, so

$$k = a_1 + a_4 + v + e - 3$$

and $a_4 = a_3 + 2$. Two cases arise. If $a_1 = 1$, then $a_2 = 2, a_3 = 3, a_4 = 5$. Also, a_5 cannot equal 6, because that would imply $a_1 + a_5 = 7 = a_2 + a_4$, contradicting the well-spread property. Every integer up to $v + e$ must occur as a label, so $b_1 = 4$ and $b_2 = 6$. So (2.14) is $a_{v-1} - a_{v-2} = b_2 - b_1 = 2$. But $a_4 - a_3 = 2$, so $d_{v-1,v-2} = d_{34}$, in contradiction of Lemma 2.9. In the other case, $a_1 = 2$, we obtain $a_2 = 3, a_3 = 4, a_4 = 6$, so $b_1 = 1, b_2 = 5$, and $a_{v-1} - a_{v-2} = 4 = a_4 - a_1$, again a contradiction.

If $a_1 + a_4$ is the next-smallest difference, we have

$$k = a_1 + a_4 + v + e - 2,$$

so $a_4 = a_3 + 1$. If $a_1 = 1$ and $a_2 = 3$, it is easy to see that $b_1 = 2, b_2 = 6$, and we get the contradiction $a_{v-1} - a_{v-2} = a_4 - a_1 = 4$. Otherwise $a_2 \geq 4$, so 3 is an edge-label. If $a_1 = 1$, then $b_1 = 2, b_2 = 3$, and $a_{v-1} - a_{v-2} = 1 = a_3 - a_2$. If $a_1 = 2$, then $b_1 = 1, b_2 = 3$, and $a_{v-1} - a_{v-2} = 2 = a_4 - a_2$. In every case, a contradiction is obtained. So we have

Theorem 2.13 *The complete graph K_v does not have an edge-magic total labeling if $v > 6$.*

This theorem was first proven in [55] (see also [54]) but the above proof follows that in [22].

2.3.3 All Edge-Magic Total Labelings of Complete Graphs

The proof in the preceding section used the fact that $v > 7$. There are edge-magic total labelings for all smaller complete graphs except K_4 and K_7 (which were excluded by Corollarys 2.1.1 and 2.12.1). A complete search has been made for smaller orders, and we now list all edge-magic total labelings for complete graphs.

In each case we list the possible values for the magic sum k, the corresponding sum of vertex labels s, and the set S of vertex labels that realize that value s and give an edge-magic total labeling.

The values to be considered are determined by equations (2.2) and (2.4). For $K_v, e = \binom{v}{2}$. So (2.2) becomes $\frac{1}{2}v(v+1) \leq s \leq \frac{1}{2}v(v^2+1)$, while (2.4) is

$$k = \frac{v(v+1)(v^2+v+2)+8(v-2)s}{4v(v-1)}.$$

For example, when $v = 6$, we have $21 \leq s \leq 111$ and $k = \frac{1}{15}(231 + 4s)$. As k is an integer, $s \equiv 6 \pmod{15}$, and the possibilities are $s = 21, 36, 51, 66, 81, 96, 111, k = 21, 25, 29, 33, 37, 41, 45$.

- K_2 is trivially possible. Label 1, 2, or 3 can be given to the edge; in each case $k = 6$.

- K_3 The magic sums to be considered are $k = 9, 10, 11, 12$.

 $k = 9, s = 6, S = \{1,2,3\}.$
 $k = 10, s = 9, S = \{1,3,5\}.$
 $k = 11, s = 12, S = \{2,4,6\}.$
 $k = 12, s = 15, S = \{4,5,6\}.$

- K_4 No solutions, by Corollary 2.1.1.

- K_5 The magic sums to be considered are $k = 18, 21, 24, 27, 30$. Theorem 2.12 tells us that no solutions exist when k is odd, so only 18, 24 and 30 are listed.

 $k = 18, s = 20, S = \{1, 2, 3, 5, 9\}$.
 $k = 24, s = 40, S = \{1, 8, 9, 10, 12\}$.
 $k = 24, s = 40, S = \{4, 6, 7, 8, 15\}$.
 $k = 30, s = 60, S = \{7, 11, 13, 14, 15\}$.

- K_6 The magic sums to be considered are $k = 21, 25, 29, 33, 37, 41, 45$.

 $k = 21, s = \ \ 21$, no solutions.
 $k = 25, s = \ \ 36, S = \{1, 3, 4, 5, 9, 14\}$.
 $k = 29, s = \ \ 51, S = \{2, 6, 7, 8, 10, 18\}$.
 $k = 33, s = \ \ 66$, no solutions.
 $k = 37, s = \ \ 81, S = \{4, 12, 14, 15, 16, 20\}$.
 $k = 41, s = \ \ 96, S = \{8, 11, 17, 18, 19, 21\}$.
 $k = 45, s = 111$, no solutions.

From this it follows that the magic strengths of K_2, K_3, K_5, and K_6 are 6, 9, 18, and 25, respectively.

2.3.4 Complete Subgraphs

If $A = (a_1, a_2, \ldots, a_n)$ is any Sidon sequence of length n, we define

$$\sigma(A) = a_n - a_1 + 1$$
$$\rho(A) = a_n + a_{n-1} - a_2 - a_1 + 1$$
$$= \sigma(A) + a_{n-1} - a_2$$
$$\sigma^*(n) = \min \sigma(A)$$
$$\rho^*(n) = \min \rho(A)$$

where the minima are taken over all Sidon sequences A of length n. σ is called the *size* of the sequence. Without loss of generality one can assume $a_1 = 1$ when constructing a sequence, and then the size equals the largest element.

We now revert to the more general case, where the graph G has an edge-magic total labeling λ and G contains a complete subgraph H on n vertices. x_1, x_2, \ldots, x_n are the vertices of H, and $a_i = \lambda(x_i)$. We assume $a_1 < a_2 < \cdots < a_n$, so $A = (a_1, a_2, \ldots, a_n)$ is a Sidon sequence of length n. Then

$$\lambda(x_n x_{n-1}) = k - a_n - a_{n-1},$$

and since $\lambda(x_n x_{n-1})$ is a label,

$$k - a_n - a_{n-1} \geq 1. \tag{2.16}$$

Similarly

$$\lambda(x_2 x_1) = k - a_2 - a_1,$$

and since $\lambda(x_2 x_1)$ is a label,

$$k - a_2 - a_1 \leq v + e. \tag{2.17}$$

Combining (2.16) and (2.17) we have

$$v + e \geq a_n + a_{n-1} - a_2 - a_1 + 1 = \rho(A) \geq \rho^*(n).$$

Theorem 2.14 [55] *If the edge-magic graph G contains a complete subgraph with n vertices, then the number of vertices and edges in G is at least $\rho^*(n)$.*

Exercise 2.5 *Suppose $G = K_n \cup tK_1$. In other words, G consists of a K_n together with t isolated vertices. Prove that if G is edge-magic, then*

$$t \geq \rho^*(n) - n - \binom{n}{2}.$$

The *magic number* $M(n)$ of K_n is defined to be the smallest t such that $K_n \cup tK_1$ is edge-magic. So the preceding exercise shows that

$$M(n) \geq \rho^*(n) - n - \binom{n}{2}.$$

Evaluation, bounds

In view of the above theorem, it is interesting to know more about Sidon sequences. Some bounds on $\sigma^*(n)$ and $\rho^*(n)$ are known:

Theorem 2.15 [53] $\sigma^*(n) \geq 4 + \binom{n-1}{2}$ *when $n \geq 7$.*

The proof appears in [53].

Theorem 2.16 [53] $\rho^*(n) \geq 2\sigma^*(n-1)$ *when $n \geq 4$.*

Proof. Consider the sequences

$$A = (a_1, a_2, \ldots, a_n)$$
$$B = (a_1, a_2, \ldots, a_{n-1})$$
$$C = (a_2, a_3, \ldots, a_n)$$

where $n \geq 4$. Clearly

$$\rho^*(n) \geq \rho(A)$$
$$= a_n + a_{n-1} - a_2 - a_1 + 1$$
$$= (a_n - a_2 + 1) + (a_{n-1} - a_1 + 1) - 1$$
$$= \sigma(B) + \sigma(C) - 1$$
$$\geq 2\sigma^*(n-1) - 1.$$

Moreover, equality can apply only if $\sigma(B) = \sigma(C) = \sigma^*(n-1)$. But

$$\sigma(B) = \sigma(C) \Rightarrow a_n - a_2 = a_{n-1} - a_1$$
$$\Rightarrow a_{n-1} + a_2 = a_n + a_1,$$

which is impossible for a Sidon sequence A. Since σ^* and ρ^* are integral,

$$\rho^*(n) \geq 2\sigma^*(n-1).$$ □

Exercise 2.6 *Prove that, when $n \geq 7$,*

$$\rho^*(n) \geq n^2 - 5n + 14. \tag{2.18}$$

From Theorems 2.16 and 2.15, $\rho^*(n) \geq 2\sigma^*(n-1) \geq 2(4 + \binom{n-2}{2}) = 8 + (n-2)(n-3) = n^2 - 5n + 14$.

In practice, values of $\sigma^*(n)$ and $\rho^*(n)$ have been calculated using an exhaustive, backtracking approach. The following result proves helpful in restricting the search for $\rho^*(n)$, once some σ^* values are known.

Theorem 2.17 *Suppose the sequence $A = (1, x, \ldots, y, z)$ satisfies $\rho(A) = \rho^*(n)$, and suppose B is any sequence for which $\rho(B)$ is known. Then*

$$\sigma^*(n) \leq z \leq \rho(B) - \sigma^*(n-2) + 1$$

and

$$x \leq (\rho(B) - \sigma^*(n-2) + 1) - \sigma^*(n-1) + 1.$$

Proof. Since (x, \ldots, y) is a Sidon sequence,

$$y - x + 1 \geq \sigma^*(n-2).$$

But

$$\rho^*(n) = z + y - x$$

so

$$z = \rho^*(n) - (y - x)$$
$$\leq \rho^*(n) - \sigma^*(n-2) + 1$$
$$\leq \rho(B) - \sigma^*(n-2) + 1.$$

Also (x, \ldots, y, z) is Sidon, so

$$z - x \geq \sigma^*(n-1) - 1$$

and the second part of the theorem follows from the upper bound for z. □

We know the following small values of the two functions. The values for $n \leq 8$ are calculated in [53] and listed in [55].

$$
\begin{array}{ll}
\sigma^*(3) = 3 & \rho^*(3) = 3 \\
\sigma^*(4) = 5 & \rho^*(4) = 6 \\
\sigma^*(5) = 8 & \rho^*(5) = 11 \\
\sigma^*(6) = 13 & \rho^*(6) = 19 \\
\sigma^*(7) = 19 & \rho^*(7) = 30 \\
\sigma^*(8) = 25 & \rho^*(8) = 43 \\
\sigma^*(9) = 35 & \rho^*(9) = 62 \\
\sigma^*(10) = 46 & \rho^*(10) = 80 \\
\sigma^*(11) = 58 & \rho^*(11) = 110 \\
\sigma^*(12) = 72 & \rho^*(12) = 137
\end{array}
$$

Sample sequences attaining the σ^* values are:

$\sigma^*(1)$ through $\sigma^*(6)$: 1 2 3 5 8 13 (or part thereof);

$\sigma^*(7)$: 1 2 3 5 9 14 19;

$\sigma^*(8)$: 1 2 3 5 9 15 20 25;

$\sigma^*(9)$: 1 2 3 5 9 16 25 30 35;

$\sigma^*(10)$: 1 2 8 11 14 22 27 42 44 46;

$\sigma^*(11)$: 1 2 6 10 18 32 34 45 52 55 58;

$\sigma^*(12)$: 1 2 3 8 13 23 38 41 55 64 68 72.

The same sequences attain $\rho^*(n)$ for $n, = 1, 2, 3, 4, 5, 6, 8$. For the other values, examples are

$\rho^*(7)$: 1 6 8 10 11 14 22;

$\rho^*(9)$: 1 5 7 9 12 17 26 27 40;

$\rho^*(10)$: 1 2 3 5 9 16 25 30 35 47;

$\rho^*(11)$: 1 2 3 5 9 16 25 30 35 47 65.

$\rho^*(12)$: 1 3 5 8 11 21 30 39 51 62 63 77.

Note: The only other sequence of length 7 with $\rho = 30$ is 1, 9, 12, 13, 15, 17, 22.

From Theorem 2.17, we have

For $n = 7$	$x \leq 12$	$19 \leq z \leq 24$
For $n = 8$	$x \leq 13$	$25 \leq z \leq 31$
For $n = 9$	$x \leq 21$	$35 \leq z \leq 45$
For $n = 10$	$x \leq 30$	$46 \leq z \leq 64$
For $n = 11$	$x \leq 32$	$58 \leq z \leq 77$
For $n = 12$	$x \leq 36$	$72 \leq z \leq 93$

and these bounds were used in calculating the example sequences for $\rho^*(n)$ when $n \geq 7$.

A greedy approach

Here is a simple observation. If $(a_1, a_2, \ldots, a_{n-1})$ is well spread, then none of its sums can exceed $a_{n-2} + a_{n-1}$. Put $a_n = a_{n-1} + a_{n-2}$. Then all the sums $a_i + a_n$ are new, and (since the sequence is strictly monotonic) they are all different. So we have a new well-spread sequence. This will be useful in constructing the smallest well-spread sequences for small orders: for example, after observing that $(1, 2, 3, 5, 8, 13)$ is a minimal example for $n = 6$, one need not test any sequence in the case $n = 7$ which has size greater than 21. (Unfortunately, 21 is not small enough.)

Suppose $(a_1, a_2, \ldots, a_{n-1})$ had minimal size, and put $a_1 = 1$. Then $\sigma^*(n-1) = a_{n-1}$. Clearly $a_{n-2} < a_{n-1}$, so we have the (bad) bound

$$\sigma^*(n) \leq 2\sigma^*(n-1) - 1.$$

Another application of this idea comes from noticing that the recursive construction $a_1 = 1, a_2 = 2, a_n = a_{n-1} + a_{n-2}$ gives a well-spread sequence. This is the Fibonacci sequence, except that the standard notation for the Fibonacci numbers has $f_1 = f_2 = 1, f_3 = 2$, etc. So we have a well-spread sequence with its size equal to the $(n+1)$-th term of the Fibonacci sequence: $a_n = f_{n+1}$. Therefore

$$\sigma^*(n) \leq \frac{1}{\sqrt{5}} \left(\frac{1 + \sqrt{5}}{2} \right)^{n+1} - \frac{1}{\sqrt{5}} \left(\frac{1 - \sqrt{5}}{2} \right)^{n+1}.$$

The same reasoning shows that

$$\rho^*(n) \leq f_{n+1} + f_n - 2 = f_{n+2} - 2.$$

Note. For further information on the Fibonacci numbers, see, for example, Sect. 7.1 of [19].

Exercise 2.7 (Continuation of Exercise 2.5.) *The **magic number** $M(n)$ of K_n was defined to be the smallest t such that $K_n \cup tK_1$ is edge-magic. We know that*

$M(n) \geq \rho^*(n) - n - \binom{n}{2}$. *Find an upper bound for $M(n)$. (It need not be a good upper bound. The point is to show that* some *upper bound exists.)*

Research Problem 2.2 (Continuation of Exercise 2.7.) *Find $M(7)$. Find $M(8)$.*

Research Problem 2.3 *Find good lower and upper bounds on $M(n)$, as a function of n.*

Exercise 2.8 *Use the results on the evaluation of Sidon sequences to prove Theorem 2.13. (Note: this was the original proof of Theorem 2.13, as given in [54, 55].)*

2.3.5 Split Graphs

A *split graph* $K_{m+n \backslash m}$ consists of a complete graph K_n, a null graph \overline{K}_m, and the $K_{m,n}$ joining the two vertex-sets.

Suppose $K_{m+n \backslash m}$ is edge-magic. This graph has $v = m + n$, $e = mn + \binom{n}{2}$, and $v + e = m + n + mn + \binom{n}{2} = \frac{1}{2}(2m + n)(n + 1)$, so by Theorem 2.14

$$\frac{1}{2}(2m + n)(n + 1) \geq \rho^*(n).$$

This gives no useful information for small n. However, provided $n > 6$, (2.18) applies and from

$$\frac{1}{2}(2m + n)(n + 1) \geq n^2 - 5n + 14$$

we find that there is no edge-magic total labeling of $K_{m+n \backslash m}$ unless

$$m \geq \frac{n - 12}{2} + \frac{20}{n + 1}. \tag{2.19}$$

Research Problem 2.4 *Observe that $K_{2n \backslash n}$ always satisfies (2.19). Does it always have an edge-magic total labeling?*

2.4 Cycles

The cycle C_v is regular of degree 2 and has v edges. So (2.2) becomes

$$v(v + 1) \leq 2s \leq 2v^2 + v(v + 1) = v(3v + 1),$$

and (2.4) is

$$kv = s + v(2v + 1),$$

whence v divides s; in fact $s = (k - 2v - 1)v$. When v is odd, s has $v + 1$ possible values $\frac{1}{2}v(v+1), \frac{1}{2}v(v+3), \ldots, \frac{1}{2}v(v+2i-1), \ldots, \frac{1}{2}v(3v+1)$, with corresponding magic sums $\frac{1}{2}(5v+3), \frac{1}{2}(5v+5), \ldots, \frac{1}{2}(5v+2i+1), \ldots, \frac{1}{2}(7v+3)$. For even v, there are v values $s = \frac{1}{2}v^2 + v, \frac{1}{2}v^2 + 2v, \ldots, \frac{1}{2}v^2 + iv, \ldots, \frac{3}{2}v^2$, with corresponding magic sums $\frac{5}{2}v + 2, \frac{5}{2}v + 3, \ldots, \frac{5}{2}v + i + 1, \ldots, \frac{7}{2}v + 1$.

Kotzig and Rosa [54] proved that all cycles are edge-magic, producing examples with $k = 3v + 1$ for v odd, $k = \frac{5}{2}v + 2$ for $v \equiv 2 \pmod 4$ and $k = 3v$ for $v \equiv 0 \pmod 4$. In [33], labelings are exhibited for the minimum values of k in all cases. For convenience we give proofs for all cases, not exactly the same as the proofs in the papers cited. In each case the proof consists of exhibiting a labeling. If vertex-names need to be cited, we assume the cycle to be (u_1, u_2, \ldots, u_v).

Theorem 2.18 *If v is odd, then C_v has an edge-magic total labeling with $k = \frac{1}{2}(5v + 3)$.*

Proof. Say $v = 2m + 1$. Consider the cyclic vertex labeling $(1, m + 1, 2m + 1, m, \ldots, m + 2)$, where each label is derived from the preceding one by adding $m(\bmod\ 2m+1)$. The successive pairs of vertices have sums $m+2, 3m+2, 3m+1, \ldots, m + 3$, which are all different. If $k = 5m + 4$, the edge labels are $4m + 2, 2m+2, 2m+3, \ldots, 4m+1$, as required. We have an edge-magic total labeling with $k = 5m + 4 = \frac{1}{2}(5v + 3)$ and $s = \frac{1}{2}v(v + 1)$ (the smallest possible values). \square

By duality, we have:

Corollary 2.18.1 *Every odd cycle has an edge-magic total labeling with $k = \frac{1}{2}(7v + 3)$.*

Theorem 2.19 *Every odd cycle has an edge-magic total labeling with $k = 3v+1$.*

Proof. Again write $v = 2m + 1$. Consider the cyclic vertex labeling $(1, 2m + 1, 4m + 1, 2m - 1, \ldots, 2m + 3)$; in this case each label is derived from the preceding one by adding $2m(\bmod\ 4m+2)$. The construction is such that the second, fourth, $\ldots, 2m$-th vertices receive labels between 2 and $2m + 1$ inclusive, while the third, fifth, $\ldots, (2m + 1)$-th receive labels between $2m + 2$ and $4m + 1$. The successive pairs of vertices have sums $2m + 2, 6m + 2, 6m, 6m - 2, \ldots, 2m + 4$; if $k = 3v + 1 = 6m + 4$, the edge labels are $4m + 2, 2, 4, \ldots, 4m$. We have an edge-magic total labeling with $k = 3v + 1$ and $s = v^2$ (the case $i = \frac{1}{2}(v + 1)$ in the list). \square

Corollary 2.19.1 *Every odd cycle has an edge-magic total labeling with $k = 3v + 2$.*

Figure 2.3 shows examples with $v = 7$ of the constructions in Theorems 2.18 and 2.19; they have $k = 19$ and 22, respectively. (Only the vertex labels are shown in the figure; the edge labels can be found by subtraction.)

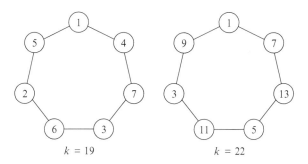

$$k = 19 \qquad\qquad k = 22$$

Fig. 2.3. Two edge-magic total labelings of C_7

Theorem 2.20 *If v is even, then C_v has an edge-magic total labeling with $k = \frac{1}{2}(5v + 4)$.*

Proof. Write $v = 2m$. If m is even,

$$\lambda(u_i) = \begin{cases} (i+1)/2 & \text{for } i = 1, 3, \ldots, m+1 \\ 3m & \text{for } i = 2 \\ (2m+i)/2 & \text{for } i = 4, 6, \ldots, m \\ (i+2)/2 & \text{for } i = m+2, m+4, \ldots, 2m \\ (2m+i-1)/2 & \text{for } i = m+3, m+5, \ldots, 2m-1, \end{cases}$$

while if m is odd,

$$\lambda(u_i) = \begin{cases} (i+1)/2 & \text{for } i = 1, 3, \ldots, m \\ 3m & \text{for } i = 2 \\ (2m+i+2)/2 & \text{for } i = 4, 6, \ldots, m-1 \\ (m+3)/2 & \text{for } i = m+1 \\ (i+3)/2 & \text{for } i = m+2, m+4, \ldots, 2m-1 \\ (2m+i)/2 & \text{for } i = m+3, m+5, \ldots, 2m-2 \\ m+2 & \text{for } i = 2m. \end{cases}$$

\square

From Theorems 2.2, 2.18 and 2.20, it follows that the magic strength of a cycle of length v is $\frac{1}{2}\lceil(5v + 3)\rceil$.

Corollary 2.20.1 *Every cycle of even length has an edge-magic total labeling with $k = \frac{1}{2}(7v + 2)$.*

In Fig. 2.4 we present two examples, for $v = 8$ and 10. The constructions of Theorem 2.20 yield $k = 22$ and 27.

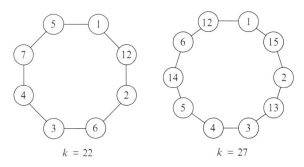

$$k = 22 \qquad\qquad k = 27$$

Fig. 2.4. Edge-magic total labelings of C_8 and C_{10}

Theorem 2.21 *Every cycle of length divisible by 4 has an edge-magic total labeling with $k = 3v$.*

Proof. For $v = 4$ the result is given by Theorem 2.20. So assume $v \geq 8$, write $v = 4m, m > 1$. The required labeling is

$$\lambda(u_i) = \begin{cases} i & \text{for } i = 1, 3, \ldots, 2m - 1 \\ 4m + i + 1 & \text{for } i = 2, 4, \ldots, 2m - 2 \\ i + 1 & \text{for } i = 2m, 2m + 2, \ldots, 4m - 2 \\ 4m + i & \text{for } i = 2m + 1, 2m + 3, \ldots, 4m - 3 \\ 2 & \text{for } i = 4m - 1 \\ 2v - 2 & \text{for } i = 4m. \end{cases}$$

\square

Corollary 2.21.1 *Every cycle of length divisible by 4 has an edge-magic total labeling with $k = 3v + 3$.*

Research Problem 2.5 [33] *If v is odd, does C_v have an edge-magic total labeling for every magic sum k satisfying $\frac{1}{2}(5v + 3) \leq k \leq \frac{1}{2}(7v + 3)$? If v is even, does C_v have an edge-magic total labeling for every magic sum k satisfying $\frac{5}{2}v + 2 \leq k \leq \frac{7}{2}v + 1$?*

2.4.1 Small Cycles

We list all edge-magic total labelings of cycles up to C_6.

There are four labelings of C_3: see under K_3, in Sect. 2.3.3.

For C_4, the possibilities are $k = 12, 13, 14, 15$, with $s = 12, 16, 20, 24$, respectively. The unique solution for $k = 12$ is the cyclic vertex-labeling $(1, 3, 2, 6)$. For $k = 13$ there are two solutions: $(1, 5, 2, 8)$ and $(1, 4, 6, 5)$. The other cases are duals of these two.

For C_5, one must consider $k = 14, 15, 16$ ($s = 15, 20, 25$) and their duals. The unique solution for $k = 14$ is $(1, 4, 2, 5, 3)$ (the solution from Theorem 2.18). There are no solutions for $k = 15$. For $k = 16$, one obtains $(1, 5, 9, 3, 7)$ (the solution from Theorem 2.19) and also $(1, 7, 3, 4, 10)$. Many other possible sets S must be considered when $k = 15$ or 16, but all can be eliminated using the following observation. The set S cannot contain three labels that add to k: for, in C_5, some pair of the corresponding vertices must be adjacent (given any three vertices of C_5, at least two must be adjacent), and the edge joining them would require the third label.

C_6 has possible sums $k = 17, 18, 19$ ($s = 24, 30, 36$) and duals. For $k = 17$ there are three solutions: $(1, 5, 2, 3, 6, 7)$, $(1, 6, 7, 2, 3, 5)$, and $(1, 5, 4, 3, 2, 9)$. Notice that two non-isomorphic solutions have the same set of vertex labels. There is one solution for $k = 18$, $(1, 8, 4, 2, 5, 10)$, and six for $k = 19$, namely $(1, 6, 11, 3, 7, 8), (1, 7, 3, 12, 5, 8), (1, 8, 7, 3, 5, 12), (1, 8, 9, 4, 3, 11), (2, 7, 11, 3, 4, 9)$, and $(3, 4, 5, 6, 11, 7)$.

In the case of C_7, the possible magic sums run from 19 to 26, and Godbold and Slater [33] found that all can be realized; there are 118 labelings up to isomorphism. The corresponding numbers for C_8, C_9, and C_{10} are 282, 1540 and 7092, [33].

2.4.2 Generalizations of Cycles

Paths

The path P_n can be viewed as a cycle C_n with an edge deleted.

Say λ is an edge-magic total labeling of C_n with the property that label $2n$ appears on an edge. If that edge is deleted, the result is a P_n with an edge-magic total labeling.

For every n, there is an edge-magic total labeling of C_n in which $2n$ appears on an edge—the labelings in Theorems 2.18 and 2.20 have this property. Deleting this edge yields a path, on which the labeling is edge-magic. So:

Theorem 2.22 *All paths have edge-magic total labelings.*

However, this simple method does not yield the smallest possible magic sum. Avadayappan et al. [5] give the following simple construction for a super edge-magic labeling of a path. Say $n = 2m$ or $2m+1$. Label the vertices x_1, x_2, \ldots, x_{2m}, preceded by x_0 if n is odd. Vertices $x_1, x_3, x_5, \ldots, x_{2n-1}$ receive labels 1, 2, 3, \ldots, n, while even vertices ($x_0, x_2, x_4, \ldots, x_{2n}$ when n is odd, $x_2, x_4, x_6, \ldots, x_{2n}$ in the even case) receive $n+1, n+2, \ldots$ in order. When this labeling is completed in the obvious way, the magic sum is $\lceil \frac{1}{2}(5n + 1) \rceil$, the theoretical minimum. So the magic strength of P_n is $\lceil \frac{1}{2}(5n + 1) \rceil$.

Exercise 2.9 *Prove that the minimum possible magic sum for the path P_n is* $\left\lceil \frac{1}{2}(5n + 1) \right\rceil$.

Suns

An n-sun is a cycle C_n with an edge terminating in a vertex of degree 1 attached to each vertex. A 4-sun is shown in Figure 2.5.

Theorem 2.23 *All suns are edge-magic.*

Proof. First we treat the odd case. Denote by λ the edge-magic total labeling of C_n given in Theorem 2.18. We construct a labeling μ which has $\mu(u) = \lambda(u) + n$ whenever u is a vertex or edge of the cycle. If a vertex has label x, then the new vertex attached to it has label a_x, where $a_x \equiv x - \frac{1}{2}(n - 1) \pmod{n}$ and $1 \le a_x \le n$, and the edge joining them has label b_x, where $b_x \equiv n + 1 - 2x \pmod{n}$ and $3n + 1 \le b_x \le 4n$. Then μ is an edge-magic total labeling with $k = \frac{1}{2}(11n + 3)$.

In the even case, λ is the edge-magic total labeling of C_n given in Theorem 2.20. The labeling μ again has $\mu(u) = \lambda(u) + n$ whenever u is an element of the cycle. The vertex with label x is adjacent to a new vertex with label a_x, and the edge joining them has label b_x, where:

- If $1 \le x \le \frac{1}{2}n$, then
 - $a_x \equiv x + \frac{1}{2}n \pmod{n}$ and $1 \le a_x \le n$,
 - $b_x \equiv 2 - 2x \pmod{n}$ and $3n + 1 \le b_x \le 4n$;
- If $1 + \frac{1}{2}n \le x < n$ then
 - $a_x \equiv x + \frac{1}{2}n + 1 \pmod{n}$ and $1 \le a_x \le n$,
 - $b_x \equiv 1 - 2x \pmod{n}$ and $3n + 1 \le b_x \le 4n$;
- $a_{\frac{3n}{2}} = b_{\frac{3n}{2}} = 1$.

Then μ is an edge-magic total labeling with $k = \frac{1}{2}(11n + 4)$. □

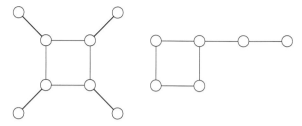

Fig. 2.5. A 4-sun and a $(4, 2)$-kite

Kites

An (n, t)-kite consists of a cycle of length n with a t-edge path (the *tail*) attached to one vertex. A (4,2) kite is shown in Figure 2.5. We write its labeling as the list of labels for the cycle (ending on the attachment point), separated by a semicolon from the list of labels for the path (starting at the vertex nearest the cycle).

Theorem 2.24 *An $(n, 1)$-kite (a kite with tail length 1) is edge-magic.*

Proof. For convenience, suppose the tail vertex is y and its point of attachment is z.

First, suppose n is odd. Denote by λ the edge-magic total labeling of C_n given in Theorem 2.18, with the vertices arranged so that $\lambda(z) = \frac{1}{2}(n + 1)$. Define a labeling μ by $\mu(x) = \lambda(x) + 1$ whenever x is a vertex or edge of the cycle, $\mu(y) = 2n + 2$ and $\mu(y, z) = 1$. Then μ is an edge-magic total labeling with $k = \frac{1}{2}(5n + 9)$.

If n is even, λ is the edge-magic total labeling of Theorem 2.20, with $\lambda(z) = \frac{1}{2}(n + 2)$. Define a labeling μ by $\mu(x) = \lambda(x) + 1$ whenever x is a vertex or edge of the cycle, $\mu(y) = 2n + 2$ and $\mu(y, z) = 1$. Then μ is an edge-magic total labeling with $k = \frac{1}{2}(5n + 10)$. ☐

Park et al. proved that an (n, t)-kite is super edge-magic if and only if n is even (see Sect. 2.9 below).

Research Problem 2.6 *Investigate the edge-magic properties of (n, t)-kites when n is odd, for general t.*

Exercise 2.10 *A triangular book $B_{3,n}$ consists of n triangles with a common edge. Prove that all triangular books are edge-magic.*

Research Problem 2.7 *The k-cycle book $B_{k,n}$ consists of n copies of C_k with a common edge. Are all k-cycle books edge-magic?*

Research Problem 2.8 *The books described in the preceding exercise and problem can be generalized by replacing the common edge by a path. Investigate the edge-magic properties of these graphs.*

2.5 Complete Bipartite Graphs

An edge-magic total labeling of a complete bipartite graph can be specified by giving two sets S_1 and S_2 of vertex labels.

Theorem 2.25 [54] *The complete bipartite graph $K_{m,n}$ is magic for any m and n.*

Proof. The sets $S_1 = \{n+1, 2n+2, \ldots, m(n+1)\}$, $S_2 = \{1, 2, \ldots, n\}$, define an edge-magic total labeling with $k = (m+2)(n+1)$. □

Research Problem 2.9 *Recall that the complete tripartite graph $K_{m,n,p}$ has three sets of vertices, of sizes m, n, p. Does $K_{m,n,p}$ always have an edge-magic total labeling?*

Research Problem 2.10 *Generalize Research Problem 2.9 to complete t-partite graphs (t parts).*

2.5.1 Small Cases

A computer search has been carried out for edge-magic total labelings of $K_{2,3}$. The usual considerations show that $14 \le k \le 22$, with cases $k = 19, 20, 21, 22$ being the duals of cases $k = 17, 16, 15, 14$. The solutions up to $k = 18$ are

$$k = 14, \text{ no solutions}$$
$$k = 15, \ S_1 = \{1, 2\}, \quad S_2 = \{3, 6, 9\}$$
$$k = 16, \ S_1 = \{1, 2\}, \quad S_2 = \{5, 8, 11\}$$
$$S_1 = \{1, 3\}, \quad S_2 = \{5, 6, 11\}$$
$$S_1 = \{4, 6\}, \quad S_2 = \{1, 2, 7\}$$
$$S_1 = \{4, 8\}, \quad S_2 = \{1, 2, 3\}$$
$$k = 17, \ S_1 = \{1, 8\}, \quad S_2 = \{5, 6, 7\}$$
$$S_1 = \{5, 6\}, \quad S_2 = \{1, 4, 9\}$$
$$k = 18, \ S_1 = \{1, 5\}, \quad S_2 = \{9, 10, 11\}$$
$$S_1 = \{7, 11\}, S_2 = \{1, 2, 3\}.$$

(The last two are of course duals.)

For $K_{3,3}$ one has $18 \le k \le 30$, and k must be even. Cases $k = 26, 28, 30$ are dual to cases $k = 22, 20, 18$. The solutions are

$$
\begin{aligned}
k &= 18, \text{ no solutions} \\
k &= 20, \ S_1 = \{1, 2, 3\}, \quad S_2 = \{4, 8, 12\} \\
&\qquad\ \ S_1 = \{1, 2, 9\}, \quad S_2 = \{4, 6, 8\} \\
k &= 22, \ S_1 = \{1, 2, 3\}, \quad S_2 = \{7, 11, 15\} \\
&\qquad\ \ S_1 = \{1, 3, 5\}, \quad S_2 = \{7, 8, 15\} \\
&\qquad\ \ S_1 = \{1, 5, 12\}, \ S_2 = \{6, 7, 8\} \\
k &= 24, \text{ no solutions.}
\end{aligned}
$$

2.5.2 Stars

Lemma 2.26 *In any edge-magic total labeling of a star, the center receives label* 1, $n + 1$ *or* $2n + 1$.

Proof. Suppose the center receives label x. Then

$$
kn = \binom{2n + 2}{2} + (n - 1)x. \tag{2.20}
$$

Reducing (2.20) modulo n we find

$$
x \equiv (n + 1)(2n + 1) \equiv 1
$$

and the result follows. $\qquad\square$

Theorem 2.27 *There are* $3 \cdot 2^n$ *edge-magic total labelings of* $K_{1,n}$, *up to equivalence.*

Proof. Denote the center of a $K_{1,n}$ by c, the leaves by v_1, v_2, \ldots, v_n and edge (c, v_i) by e_i. From Lemma 2.26 and (2.20), the possible cases for an edge-magic total labeling are $\lambda(c) = 1, k = 2n + 4$, $\lambda(c) = n + 1, k = 3n + 3$ and $\lambda(c) = 2n + 1, k = 4n + 2$. As the labeling is magic, the sums $\lambda(v_i) + \lambda(e_i)$ must all be equal to $M = k - \lambda(c)$ (so $M = 2n + 3, 2n + 2$ or $2n + 1$). Then in each case there is exactly one way to partition the $2n + 1$ integers $1, 2, \ldots, 2n + 1$ into $n + 1$ sets

$$
\{\lambda(c)\}, \{a_1, b_1\}, \{a_2, b_2\}, \ldots, \{a_n, b_n\}
$$

where every $a_i + b_i = M$. For convenience, choose the labels so that $a_i < b_i$ for every i and $a_1 < a_2 < \cdots < a_n$. Then up to isomorphism, one can assume that $\{\lambda(v_i), \lambda(e_i)\} = \{a_i, b_i\}$. Each of these n equations provides two choices, according as $\lambda(v_i) = a_i$ or b_i, so each of the three values of $\lambda(c)$ gives 2^n edge-magic total labelings of $K_{1,n}$. $\qquad\square$

Exercise 2.11 *A graph is derived from a star by adding a pendant edge to one of the vertices of degree* 1. *Prove every such graph is edge-magic.*

Double Stars

The *double star* $S_{m,n}$ has two adjacent central vertices x and y. There are m leaves x_1, x_2, \ldots, x_m adjacent to x and n leaves y_1, y_2, \ldots, y_n adjacent to y. An edge-magic total labeling of this graph can be specified by the list

$$(\{\lambda(x_1), \lambda(x_2), \ldots, \lambda(x_m)\}, \lambda(x), \lambda(y), \{\lambda(y_1), \lambda(y_2), \ldots, \lambda(y_n), \}).$$

One solution for $S_{2,2}$ is $(\{8, 11\}, 2, 5, \{4, 10\})$ with $k = 16$.

Research Problem 2.11 *For what m and n does $S_{m,n}$ have an edge-magic labeling?*

2.6 Wheels

Enomoto *et al.* [25] have checked all wheels up to $n = 29$ and found that the graph is magic if $n \not\equiv 3 \pmod 4$. We shall prove that W_n is magic in all other cases.

To describe a wheel, we refer to the central vertex as the *center*, the edges adjacent to the center as *spokes*, and the other vertex on a spoke as its *terminal*. The remaining edges are *arcs*. We write c for the label on the center; the spokes receive labels a_1, a_2, \ldots, a_n, the terminal of the spoke a_i gets label b_i, and the arc from b_i to b_{i+1} is labeled s_i. (This scheme is illustrated in Fig. 2.6.) It then follows that all the sums $a_i + b_i$ in an edge-magic total labeling are the same; they equal $k - c$, where k is the magic sum.

For the cases other than $n \equiv 6 \pmod 8$, our constructions (taken from [80]) use $k = 2c$. Then $a_i + b_i = c$ for every i. It follows that c is greater than each a_i and b_i; since labels are distinct and positive, $c \geq 2n + 1$. Also $s_i = a_i + a_{i+1}$. We then need to do the following:

1. Partition the integers from 1 to $2n$ into two classes A and B so that exactly one member of $\{i, 2n + 1 - i\}$ is in each set, for every i.

2. Order the elements of A in a cyclic sequence so that the n sums of consecutive pairs comprise the integers from $2n + 2$ to $3n + 1$ in some order (or, what is equivalent, the elements of B must be ordered so that the consecutive sums are a permutation of $\{n + 1, \ldots, 2n\}$).

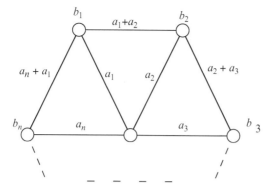

Fig. 2.6. Labeling a wheel

To visualize this process, it is perhaps easiest to consider a graph with vertices $1, 2, \ldots, 2n$, drawn in two sets so that i and $2n + 1 - i$ are vertical opposite pairs. Two vertices are connected if their sum lies between $n + 1$ and $2n$. The graph for $n = 5$ is shown in Fig. 2.7. We need to find a cycle that contains exactly one member from each opposite pair, all of whose edges have different sums. From this representation it is clear, for example, that both 1 and 2 must be terminal labels, because neither $2n$ nor $2n - 1$ have degree 2 or greater.

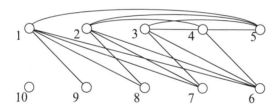

Fig. 2.7. The spoke-terminal graph

2.6.1 The Constructions for $n \not\equiv 2 \pmod 4$

We construct, for each $n \equiv 0, 1, 4$ or $5 \pmod 8$, a labeling of the terminal vertices of the wheel W_n. That is, we construct a sequence of n members of $\{1, 2, \ldots, 2n\}$ such that

(i) For each $i = 1, 2, \ldots, n$, exactly *one* of i and $2n + 1 - i$ is a member of the sequence, and

(ii) The set of sums of pairs of successive elements of the sequence (including last and first elements) is precisely $n + 1, n + 2, \ldots, 2n$.

In every case the sequence is defined in terms of several subsequences which are then joined.

It will be observed that the constructions do not provide solutions for $n = 5$ or 13. There is no sequence for $n = 5$; for $n = 13$ one example is

$$(12\ 4\ 20\ 1\ 24\ 2\ 16\ 6\ 13\ 10\ 5\ 9\ 8).$$

Case $n \equiv 0 \pmod 8$, $n \geq 8$.

For each $i \equiv 1 \pmod 4$ with $1 \leq i \leq \frac{n}{2} - 3$ define

$$S_i = (i, 2n + 1 - 3i, i + 1, 2n + 1 - 3i - 4, i + 3, 2n + 1 - 3i - 8),$$

and define

$$S = \left(\frac{n}{2} + j : j \equiv 0, 1 \pmod 4 \text{ and } 1 \leq j \leq \frac{n}{2}\right).$$

Then the desired sequence is

$$S_1 S_5 \ldots S_{\frac{n}{2}-3} S.$$

Case $n \equiv 4 \pmod 8$, $n \geq 4$.

For each $i \equiv 1 \pmod 4$ with $1 \leq i \leq \frac{n}{2} - 5$ define

$$S_i = (i + 1, 2n + 1 - 3i, i, 2n + 1 - 3i - 4, i + 3, 2n + 1 - 3i - 8),$$

and define

$$S = \left(\frac{n}{2} - 2 + j : j \equiv 0, 1 \pmod 4 \text{ and } 1 \leq j \leq \frac{n}{2} + 2\right).$$

Then the desired sequence is

$$S_1 S_5 \ldots S_{\frac{n}{2}-5} \left(\frac{n}{2}, \frac{n}{2} + 4\right) S.$$

(Note that when $n = 4$, the sequence is simply $(\frac{n}{2}, \frac{n}{2} + 4)S = (2, 6, 1, 4)$.)

Case $n \equiv 1 \pmod 8$, $n \geq 9$.

For each $i \equiv 1 \pmod 4$ with $1 \leq i \leq \frac{n-1}{2} - 3$ define

$$S_i = \{i, 2n + 1 - 3i, i + 1, 2n + 1 - 3i - 4, i + 3, 2n + 1 - 3i - 8\},$$

and define

$$S = \left\{\frac{n-1}{2} + j : j \equiv 1, 2 \pmod 4 \text{ and } 1 \leq j \leq \frac{n+1}{2}\right\}.$$

Then the desired sequence is

$$S_1 S_5 \ldots S_{\frac{n-1}{2}-3} S.$$

Case $n \equiv 5 \pmod{24}, n \geq 29$.

Define

$$S = \left(\frac{n+10}{3}, n-5, \frac{n+1}{3}, n+3, \frac{n+7}{3}, n-1, 4, 2n+1-7, 1, \right.$$
$$\left. 2n+1-3, 2, 2n+1-11 \right)$$

and

$$T = \left(\frac{n-2}{3}, n, n-3, n-4, n-7, \ldots, \frac{n+5}{2}, \frac{n-1}{2}, \frac{n+3}{2}, \frac{n+11}{2} \right).$$

For each $i \equiv 6 \pmod{8}, i \geq 6$, define

$$S_i = (i, 2n+1-(3i+1), i+2, 2n+1-(3i-3), i-1, 2n+1-(3i+5)),$$

and for each $i \equiv 1 \pmod{8}, i \geq 9$, define

$$S_i = (i, 2n+1-(3i+4), i+3, 2n+1-3i, i+1, 2n+1-(3i+8)).$$

For each $j = 1, 2, \ldots, \frac{n-29}{24}$ write

$$C_j = \left(\frac{n+10}{3} + 4j, n-5-12j, \frac{n+10}{3} + 4j - 3, \right.$$
$$\left. n-5-12j+4, \frac{n+10}{3} + 4j - 1, n-5-12j+8 \right).$$

Then the desired sequence is

$$S S_6 S_9 \ldots S_{\frac{n-35}{3}} S_{\frac{n-26}{3}} S_{\frac{n-11}{3}} T C_{\frac{n-29}{24}} \ldots C_2 C_1.$$

(Note that when $n = 29$ this sequence is $S S_6 T$.)

Case $n \equiv 21 \pmod{24}, n \geq 21$.

Define

$$S = \left(\frac{n+6}{3}, n-1, 4, 2n+1-7, 1, 2n+1-3, 2, 2n+1-11 \right)$$

and

$$T = \left(\frac{n-3}{3}, 2n+1-(n-2), \frac{n-3}{3}+2, 2n+1-(n-6), \right.$$
$$\frac{n-3}{3} - 1, n, n-3, n-4, n-7, \ldots,$$
$$\left. \frac{n+5}{2}, \frac{n-1}{2}, \frac{n+3}{2}, \frac{n+11}{2} \right).$$

For each $i \equiv 6 \pmod 8$, $i \geq 6$, and for each $i \equiv 1 \pmod 8$, $i \geq 9$, define S_i as in the $n \equiv 5 \pmod{24}$ case. Then, for each $j = 1, 2, \ldots, \frac{n-21}{24}$ write

$$C_j = \left(\frac{n+6}{3} + 4j, n - 1 - 12j, \frac{n+6}{3} + 4j - 3, n - 1 - 12j + 4, \right.$$
$$\left. \frac{n+6}{3} + 4j - 1, n - 1 - 12j + 8 \right).$$

Then the desired sequence is

$$S S_6 S_9 \ldots S_{\frac{n-27}{3}} S_{\frac{n-18}{3}} T C_{\frac{n-21}{24}} \ldots C_2 C_1.$$

(Note that when $n = 21$ this sequence is ST.)

Case $n \equiv 13 \pmod{24}$.

Define

$$S = \left(n - 1 - 8, \frac{n+5}{3} + 2, n - 1 - 4, \frac{n+5}{3} + 6, \frac{2n+1}{3}, \frac{n+2}{3}, \right.$$
$$\left. n - 1, 4, 2n + 1 - 7, 1, 2n + 1 - 3, 2, 2n + 1 - 11 \right)$$

and

$$T = \left(\frac{n+5}{3}, n, n - 3, n - 4, n - 7, \ldots, \frac{2n+1}{3} + 4, \frac{2n+1}{3} + 1, \right.$$
$$\left. \frac{2n+1}{3} - 4, \frac{2n+1}{3} - 3, \ldots, \frac{n+5}{2}, \frac{n+7}{2}, \frac{n-3}{2}, \frac{n+11}{2} \right).$$

For each $i \equiv 6 \pmod 8$, $i \geq 6$, and for each $i \equiv 1 \pmod 8$, $i \geq 9$, define S_i as in the $n \equiv 5 \pmod{24}$ case. Now we must break into two subcases:

1. $n \equiv 13 \pmod{48}$, $n \geq 61$.

Define

$$D = \left(\frac{n+5}{3} + 10, n - 1 - 20, \frac{n+5}{3} + 4, n - 1 - 16, \frac{n+5}{3} + 8, \right.$$
$$\left. n - 1 - 12, \frac{n+5}{3} + 3 \right),$$

and for each $j = 1, 2, \ldots, \frac{n-61}{48}$ write

$$C_j = \left(\frac{n+5}{3} + 10 + 8j, n - 1 - 24j - 20, \frac{n+5}{3} + 4 + 8j, \right.$$
$$n - 1 - 24j - 16, \frac{n+5}{3} + 8 + 8j, n - 1 - 24j - 12,$$
$$\frac{n+5}{3} + 6 + 8j, n - 1 - 24j - 8, \frac{n+5}{3} - 1 + 8j,$$
$$\left. n - 1 - 24j - 4, \frac{n+5}{3} + 3 + 8j, n - 1 - 24j \right).$$

Then the desired sequence is

$$SS_6S_9 \ldots S_{\frac{n-19}{3}} S_{\frac{n-10}{3}} TC_{\frac{n-61}{48}} \ldots C_2C_1D.$$

(Note that when $n = 61$ this sequence is $SS_6S_9S_14S_1TD$.)

2. $n \equiv 37 \,(\mathrm{mod}\,48), n \geq 37$.

Write $D = (\frac{n+5}{3} + 4)$, and for each $j = 1, 2, \ldots, \frac{n-37}{48}$ define

$$C_j = \left(\frac{n+5}{3} + 6 + 8j, n - 1 - 24j - 8, \frac{n+5}{3} + 8j, \right.$$

$$n - 1 - 24j - 4, \frac{n+5}{3} + 4 + 8j, n - 1 - 24j,$$

$$\frac{n+5}{3} + 2 + 8j, n - 1 - 24j + 4, \frac{n+5}{3} - 5 + 8j,$$

$$\left. n - 1 - 24j + 8, \frac{n+5}{3} - 1 + 8j, n - 1 - 24j + 12 \right).$$

Then the desired sequence is

$$SS_6S_9 \ldots S_{\frac{n-19}{3}} S_{\frac{n-10}{3}} TC_{\frac{n-37}{48}} \ldots C_2C_1D.$$

(Note that when $n = 37$ this sequence is $SS_6S_9S_14S_17TD$.)

Theorem 2.28 *Every wheel* W_n *with* $n \equiv 0$ *or* $1 \,(\mathrm{mod}\,4)$ *has an edge-magic total labeling.*

Proof. It will suffice to prove that the above constructions have the required properties (i) and (ii). We prove this for the hardest cases, the case $n \equiv 1 \,(\mathrm{mod}\,8)$ and $n \equiv 13 \,(\mathrm{mod}\,48), n \geq 61$; the proofs for $n \equiv 0$ and $4 \,(\mathrm{mod}\,8)$ are similar to the former, while the proofs in the other cases with $n \equiv 5 \,(\mathrm{mod}\,8)$ are similar to the latter.

$n \equiv 1 \,(\mathrm{mod}\,8)$

Property (i): Suppose $m \equiv 1$ or $2 \,(\mathrm{mod}\,4), 1 \leq m \leq n$. If $1 \leq m \leq \frac{n-1}{2} - 2$ then m appears in one of the S_i (as an 'i' or '$i+1$' term). If $\frac{n-1}{2} + 1 \leq m \leq n$, then m appears in S. Now suppose $m \equiv 0(\,\mathrm{mod}\,4), 4 \leq m \leq 2n-2$. If $4 \leq m \leq \frac{n-1}{2}$, then m appears in one of the S_i (as an '$i+3$' term) while if $\frac{n-1}{2}+4 \leq m \leq 2n-2$, then again m appears in one of the S_i, as follows. If $m \equiv \frac{n-1}{2}+4 \mod 12$, then m appears as a term of the form '$2n+1-3i-8$'; if $m \equiv \frac{n-1}{2}+8 \mod 12$, then m appears as a '$2n+1-3i-4$' term; if $m \equiv \frac{n-1}{2} \mod 12$ (which means $m \equiv 2n-2 \mod 12$ since $n \equiv 1 \,(\mathrm{mod}\,8)$), then m appears as a '$2n+1-3i$' term. Now since we have all $m \equiv 0 \,(\mathrm{mod}\,4), 4 \leq m \leq 2n - 2$, we also have represented all of their inverses mod $2n + 1$, which yields all of the 3 $(\mathrm{mod}\,4)$ cases between 3 and $2n - 3$. Hence we have property (i).

Property (ii): First we note that those sums from $n + 2$ to $2n - 3$ inclusive which are congruent to $3 \pmod 4$ appear as successive sums of pairs in S. Then those sums from $n + 4$ to $2n - 1$ inclusive which are congruent to $1 \bmod 4$ appear as the union of the sums of the first and second terms of each S_i, the last term in S_i with the first term in S_{i+4} and the last term in $S_{\frac{n-1}{2} - 3}$ with the first term in S. The sums from $n + 1$ to $2n$ inclusive which are congruent to $2 \bmod 4$ appear as the union of the sums of the second and third, and third and fourth terms in each S_i and the sum of the last term in S with the first term in S_1. Finally, those sums from $n + 3$ to $2n - 2$ inclusive which are congruent to $0 \bmod 4$ appear as the union of the sums of the fourth and fifth, and fifth and sixth, terms in each S_i. Hence we have property (ii).

$n \equiv 13 \pmod{48}$

The sequence for the case $n = 13$ can be verified directly, so we assume $n \geq 61$.

Property (i): Suppose $m \equiv 1$ or $2 \pmod 4$, $1 \leq m \leq n$. If $m = 1, 2, \frac{n+2}{3}$ or $\frac{2n+1}{3}$, then m appears in S. If $m = \frac{n+5}{3} + 3, \frac{n+5}{3} + 4$ or $\frac{n+5}{3} + 8$, then m appears in D. If $m = \frac{n+5}{3}$ or $\frac{n-3}{2}$, or if $\frac{n+5}{2} \leq m \leq n, m \neq \frac{2n+1}{3}$, then m appears in T. If $m \equiv 5$ or $6 \pmod 8$, $5 \leq m \leq \frac{n+5}{3} - 8$, then m appears in an S_i with $i \equiv 6 \pmod 8$, as the $i - 1$ or i term, while if $m \equiv 1$ or $2 \pmod 8$, $9 \leq m \leq \frac{n+5}{3-4}$, then m appears in an S_i with $i \equiv 1 \pmod 8$, as the i or $i + 1$ term. There remain those $m \equiv 1 \pmod 4$, $\frac{n+5}{3+7} \leq m \leq \frac{n-11}{2}$, and the $m \equiv 2 \pmod 4$, $\frac{n+5}{3+12} \leq m \leq \frac{n-1}{2}$. These terms appear in the C_js, the former as the $\frac{n+5}{3-1+8j}$ and $\frac{n+5}{3+3+8j}$ terms, $1 \leq j \leq \frac{n-61}{48}$, and the latter as the $\frac{n+5}{3+4+8j}$ and $\frac{n+5}{3+8+8j}$ terms, $1 \leq j \leq \frac{n-61}{48}$.

Now suppose $m \equiv 0 \pmod 4$, $4 \leq m \leq 2n - 2$. If $4 \leq m \leq \frac{n+5}{3+6}$ or if $n - 9 \leq m \leq 2n - 2$, then m appears in S or in one of the S_is. If $m = \frac{n+5}{3} + 10, n - 21, n - 17$ or $n - 13$, then m appears in D. There remain the cases $m \equiv 0 \pmod 4$, $\frac{n+5}{3} + 14 \leq m \leq n - 25$. Now the term $\frac{n+11}{2}$ appears in T. The terms $\frac{n+5}{3} + 14 \leq m \leq \frac{n+3}{2}$ appear in the C_js as the $\frac{n+5}{2} + 10 + 8j$ and $\frac{n+5}{3} + 6 + 8j$ terms, $1 \leq j \leq \frac{n-61}{48}$, while the terms $\frac{n+19}{2} \leq m \leq n - 25$ appear in the C_js as the $n - 1 - 24j - 4k$ terms, $0 \leq k \leq 5, 1 \leq j \leq \frac{n-61}{48}$. Since we have all $0 \pmod 4$ terms between 4 and $2n - 2$ we have also represented all $3 \pmod 4$ terms between 3 and $2n - 3$ (i.e., the inverses modulo $(2n + 1)$). This verifies property (i).

Property (ii): Sums from $n + 2$ to $2n - 3$ inclusive which are congruent to $3 \pmod 4$ appear as succesive sums of pairs of elements in T. Sums from $\frac{4n+11}{3}$ to $2n - 1$ inclusive which are congruent to $1 \pmod 4$ appear as sums in S involving 1 (i.e., $2n - 1$ and $42n - 5$) and as sums of the fourth and fifth, and fifth and sixth terms in S_i, the sixth term in S_i with the first term in S_{i+3}, and the first and second terms in S_{i+3}, where $i \equiv 6 \pmod 8$, $6 \leq i \leq \frac{n-19}{3}$. The sum

$\frac{4n-1}{3}$ appears as the sum of the sixth and seventh terms in S, while the sum $n+8$ appears as the sum of the fourth and fifth terms in S; $n+4$ appears as the sum of the last two terms in T. Now $\frac{4n-25}{3}$ and $\frac{4n-13}{3}$ appear as the sum of the last two terms in D and the sum of the last term in D with the first term in S, respectively. There remain the sums congruent to 1 (mod 4) between $n+12$ and $\frac{4n-37}{3}$ inclusive; these appear as successive sums over the eighth through twelfth terms in the C_js, $1 \leq j \leq \frac{n-61}{48}$. Sums from $\frac{4n+2}{3}$ to $2n$ inclusive which are congruent to 2 (mod 4) appear as sums of the eleventh and twelfth, and twelfth and last terms in S, the last term in S with the first term in S_6, and the first two terms in S_6; then as sums of the fourth and fifth, and fifth and sixth terms in S_i, the sixth term in S_i with the first term in $S_i + 5$, and the first and second terms in $S_i + 5$, where $i \equiv 1 \pmod 8$, $9 \leq i \leq \frac{n-10}{3-8}$; then finally as sums of the fourth and fifth, and fifth and sixth terms in $S_{\frac{n-10}{3}}$, and the sixth term in $S_{\frac{n-10}{3}}$ with the first term in T. The sum $n+1$ appears as the sum of the fifth and sixth terms in S, while the sums $\frac{4n-10}{3-4k}$, $3 \geq k \geq 0$, appear as successive sums over the second through sixth terms in D. There remain the sums congruent to 2 (mod 4) between $n+5$ and $\frac{4n-58}{3}$ inclusive; these appear as successive sums over the second through sixth terms in the C_js, $1 \leq j \leq \frac{n-61}{48}$. Sums from $\frac{4n+20}{3}$ to $2n-2$ inclusive which are congruent to 0 (mod 4) appear as sums of the eighth and ninth terms in S, and as sums of the second and third, and third and fourth terms in S_i and $S_i + 3$, where $i \equiv 6 \pmod 8$ and $6 \leq i \leq \frac{n-19}{3}$. The sum $n+3$ appears as the sum of the seventh and eighth terms in S, while the sums $\frac{4n+20}{3-4k}$, $4 \geq k \geq 1$, appear as the sum of the first two terms in D and as succesive sums over the first through fourth terms in S. There remain the sums congruent to 0 (mod 4) between $n+7$ and $\frac{4n-40}{3}$ inclusive. When $n = 61$, this sum (namely 68) appears as the sum of the last term in T (namely $\frac{n+11}{2} = 36$ and the first term in D ($\frac{n+5}{3} + 10 = 32$); when $n > 61$ these appear as the sum of the last term in T with the first term in $C_{\frac{n-61}{48}}$, the sums of the first and second, sixth and seventh, and seventh and eighth terms in each C_j, $\frac{n-61}{48} \geq j \geq 1$, the sum of the last term in C_j with the first term in $C_j - 1$, $\frac{n-61}{48} \geq j \geq 2$, and the sum of the last term in C_1 with the first term in D. This completes the verification of property (ii). □

2.6.2 The Constructions for $n \equiv 2 \pmod 4$

Case $n \equiv 6 \pmod 8$.

If $n \equiv 6 \pmod 8$, the following construction (taken from [85]) provides an edge-magic total labeling with $k = 5n + 2$.

$c = 2n.$

For $i = 1, 3, \ldots, \frac{n}{2}$,
$$\begin{cases} a_i = \frac{1}{2}(i+1) \\ b_i = 3n + 2 - \frac{1}{2}(i+1) \\ s_i = n + i. \end{cases}$$

For $i = 2, 4, \ldots, \frac{n-2}{2}$, $\begin{cases} a_i = 2n + 1 + \frac{1}{2}i \\ b_i = n + 1 - \frac{1}{2}i \\ s_i = n + i. \end{cases}$

For $i = \frac{n+4}{2}, \frac{n+8}{2}, \ldots, \frac{3n-6}{4}$, $\begin{cases} a_i = \frac{1}{2}(i + 3) \\ b_i = 3n + 1 - \frac{1}{2}(i + 1) \\ s_i = n + i + 1. \end{cases}$

For $i = \frac{n+2}{2}, \frac{n+8}{2}, \ldots, \frac{3n-2}{4}$, $\begin{cases} a_i = 2n + 1 + \frac{1}{2}i \\ b_i = n + 1 - \frac{1}{2}i \\ s_i = n + i + 1. \end{cases}$

For $i = \frac{3n+2}{4}$, $\begin{cases} a_i = \frac{1}{8}(5n + 2) \\ b_i = \frac{1}{8}(19n + 14) \\ s_i = 2n + 1. \end{cases}$

For $i = \frac{3n+6}{4}, \frac{3n+14}{4}, \ldots, n$, $\begin{cases} a_i = 2n + 2 + \frac{1}{2}i \\ b_i = n - \frac{1}{2}i \\ s_i = n + i + 1. \end{cases}$

For $i = \frac{3n+10}{4}, \frac{3n+18}{4}, \ldots, n - 3$, $\begin{cases} a_i = \frac{1}{2}(i + 5) \\ b_i = 3n - \frac{1}{2}(i + 1) \\ s_i = n + i + 1. \end{cases}$

For $i = n - 1$, $\begin{cases} a_i = \frac{1}{4}(11n + 2) \\ b_i = \frac{1}{4}(n + 6) \\ s_i = \frac{1}{4}(7n + 6). \end{cases}$

with the exceptions $\begin{cases} s_{\frac{1}{4}(3n-2)} = 2n - 1 \\ s_{n-2} = \frac{1}{4}(7n + 2) \\ s_n = \frac{1}{2}(3n + 2). \end{cases}$

Case $n \equiv 2 \,(\mathrm{mod}\ 8), n \geq 10$.

Research Problem 2.11 in our first edition was to show that wheels are edge-magic when $n \equiv 2(\mathrm{mod}\ 8)$. The following construction was given in [30]:

Let $n = 8m + 2, m \geq 1$. Thus, $v + e = 24m + 7$. Let $c = 8m + 6$ for all cases. In the case where $m = 1$ we have $(23, 2, 26, 3, 31, 4, 22, 5, 25, 8)$ as the vertex labels on the cycle with $k = 46$. When $m = 2$ we have $(38, 6, 39, 9, 40, 2, 41, 5, 42, 13, 444, 7, 45, 8, 46, 4, 55, 3)$ as the vertex labels on the cycle with $k = 78$. For $m \geq 3$ we have $k = 32m + 14$ and define

$$b_{2i-1} = \begin{cases} 16m + 5 + i & 1 \leq i \leq 3m - 1 \\ 16m + 6 + i & 3m \leq i \leq 4m \\ 24m + 7 & i = 4m + 1. \end{cases}$$

For the labels b_{2i} $(1 \leq i \leq 4m + 1)$, there are three subcases.

Subcase 1: $m \equiv 0 (\mathrm{mod}\, 3)$. Set $m = 3s$ $(s \geq 1)$. Thus, $n = 24s + 2$ and $v + e = 72s + 7$. Define

$$
b_{2i} = \begin{cases}
i + 5 & 1 \leq i \leq 6s \text{ and } m \not\equiv 0 (\mathrm{mod}\, 3) \\
i - 1 & 1 \leq i \leq 6s \text{ and } m \equiv 0 (\mathrm{mod}\, 3) \\
12s + 1 & i = 6s + 1 \\
i + 4 & 6s + 2 \leq i \leq 9s - 2 \text{ and } m \not\equiv 1 (\mathrm{mod}\, 3),\ s \geq 2 \\
i - 2 & 6s + 2 \leq i \leq 9s - 2 \text{ and } m \equiv 1 (\mathrm{mod}\, 3),\ s \geq 2 \\
15s + 3 & i = 9s - 1 \\
9s - 1 & i = 9s \\
i + 1 & 9s + 1 \leq i \leq 12s - 1 \\
4 & i = 12s \\
3 & i = 12s + 1.
\end{cases}
$$

Subcase 2: $m \equiv 1 (\mathrm{mod}\, 3)$. Set $m = 3s + 1$ $(s \geq 1)$. Thus, $n = 24s + 10$ and $v + e = 72s + 31$. Define

$$
b_{2i} = \begin{cases}
i + 5 & 1 \leq i \leq 6s \text{ and } m \not\equiv 0 (\mathrm{mod}\, 3) \\
i - 1 & 1 \leq i \leq 6s + 3 \text{ and } m \equiv 0 (\mathrm{mod}\, 3) \\
12s + 5 & i = 6s + 1 \\
6s + 5 & 6s + 2i = 6s + 2 \\
i + 2 & 6s + 4 \leq i \leq 9s + 1 \\
15s + 8 & i = 9s + 2 \\
i + 1 & 9s + 3 \leq i \leq 12s + 3 \\
4 & i = 12s + 4 \\
3 & i = 12s + 5.
\end{cases}
$$

Subcase 3: $m \equiv 2 (\mathrm{mod}\, 3)$. Set $m = 3s + 2$ $(s \geq 1)$. Thus, $n = 24s + 18$ and $v + e = 72s + 55$. Define

$$
b_{2i} = \begin{cases}
6 & i = 1 \\
i + 7 & 2 \leq i \leq 6s + 4 \text{ and } i \equiv 2 (\mathrm{mod}\, 3) \\
2 & i = 3 \\
i + 1 & 4 \leq i \leq 6s + 4 \text{ and } i \not\equiv 2 (\mathrm{mod}\, 3) \\
12s + 9 & i = 6s + 5 \\
i + 6 & 6s + 6 \leq i \leq 9s + 2 \text{ and } i \equiv 0 (\mathrm{mod}\, 3) \\
i & 6s + 6 \leq i \leq 9s + 4 \text{ and } i \not\equiv 0 (\mathrm{mod}\, 3) \\
9s + 7 & i = 9s + 3 \\
15s + 13 & i = 9s + 5 \\
9s + 5 & i = 9s + 6 \\
i + 1 & 9s + 7 \leq i \leq 12s + 7 \\
4 & i = 12s + 8 \\
3 & i = 12s + 9.
\end{cases}
$$

Exercise 2.12 *The Petersen graph P consists of two 5-cycles $x_1x_2x_3x_4x_5$ and $y_1y_3y_5y_2y_4$, together with the five edges $x_1y_1, x_2y_2, x_3y_3, x_4y_4, x_5y_5$. What is the range of possible magic sums for an edge-magic total labeling of P? Prove that P is edge-magic.*

Exercise 2.13 *A fan F_n is constructed from a wheel W_n by deleting one arc. Prove that all fans are edge-magic.*

Research Problem 2.12 *A helm H_n is constructed from a wheel W_n by adding n vertices of degree 1, one adjacent to each terminal vertex. Which helms are edge-magic?*

Research Problem 2.13 *A flower F_n is constructed from a helm H_n by joining each vertex of degree 1 to the center. Which flowers are edge-magic?*

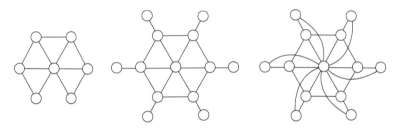

Fig. 2.8. A fan, a helm and a flower

2.7 Trees

It has been conjectured ([54], also [82]) that all trees are edge-magic. However, this seems to be a difficult problem.

We have already seen that stars and paths are edge-magic. A *caterpillar* is a graph derived from a path by hanging any number of leaves from the vertices of the path, so it can be seen as a sequence of stars where each star shares one edge with the next one. We shall show that all caterpillars are edge-magic.

The typical caterpillar is a graph $G = S^1 \cup S^2 \cup \ldots \cup S^n$, where S^i is a star with center c_i and e_i edges, and in every case S^i shares an edge with S^{i+1}. Then G has $e = \sum_1^n e_i - n + 1$ edges ($n - 1$ edges are shared by two stars), and $v = \sum_1^n e_i - n + 2$ vertices. c_i will be a leaf in S^{i-1} (unless $i = 1$) and in S^{i+1} (unless $i = n$), or equivalently the leaves of S^i will include c_{i-1} and c_{i+1}.

Theorem 2.29 [54] *All caterpillars are edge-magic.*

Proof. We describe an edge-magic total labeling λ of the caterpillar G described above. First the stars are ordered $S^1, S^3, S^5, \ldots, S^2, S^4, S^6, \ldots$ and then the leaves of the stars are labeled with the smallest positive integers, starting from 1 as a label on S^1 and ascending. When the leaves of S^i are labeled, c_{i-1} receives the smallest label (except when $i = 1$) and c_{i+1} the largest one. So the leaves of S^1 receive $1, 2, \ldots, e_1$, with $\lambda(c_2) = e_1$, then the vertices of S_3 receive $v_1, v_1 + 1, \ldots, e_1 + e_3 - 1$, with $\lambda(c_2) = e_1$ and $\lambda(c_2) = e_1 + e_3 - 1$, and so on. This uses labels $1, 2, \ldots, \sum_1^n e_i - n + 2 = v$. Then the edges are labeled. The smallest available labels, namely $v + 1, v + 2, \ldots, v + e_n$, are applied to the edges of S^n, then the next e_{n-1} to the edges of S^{n-1}, and so on until the edges of S^1 are labeled. In each star, the smallest label is given to the edge whose perimeter vertex has the largest label, and so on. The labeling is illustrated in Fig. 2.9.

Verification that the labeling has the edge-magic property is an easy exercise. □

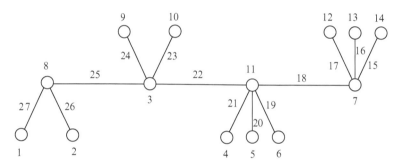

Fig. 2.9. Labeling a caterpillar

Exercise 2.14 *Verify that the labeling described in the proof of Theorem 2.9 is edge-magic.*

This exhausts our systematic knowledge of edge-magic total labelings of trees. However, Enomoto et al. [25] carried out a computer check and showed that all trees with fewer than 16 vertices are edge-magic. In fact, labelings were easy to find: they found that every tree on v vertices ($v \leq 16$) had a super edge-magic labeling.

Research Problem 2.14 *Are all trees edge-magic?*

2.8 Disconnected Graphs

2.8.1 Some Easy Cases

Kotzig and Rosa [54] showed that the one-factor F_{2n}, consisting of n independent edges, is edge-magic if and only if n is odd (see Exercise 2.15), and nK_4 is not edge-magic for n odd (see Exercise 2.1).

Exercise 2.15 *Prove that the one-factor F_{2n}, consisting of n independent edges, is edge-magic if and only if n is odd.* [54]

Another interesting type of disconnected graph is a cycle plus one disconnected edge ($K_2 \cup C_n$). In the first edition of this book, we proposed investigating whether such graphs are edge-magic. Park et al. proved something stronger:

Theorem 2.30 [50, 79] *$K_2 \cup C_n$ is super edge-magic if and only if n is even.*

The proof is lengthy, with separate analyses according to the value of n modulo 12.

Exercise 2.16 *Prove that $K_2 \cup C_3$ is not edge-magic, but $K_2 \cup C_4$ is edge-magic.*

Research Problem 2.15 *Is $K_2 \cup C_n$ edge-magic for n odd?*

2.8.2 Unions of Complete Graphs

In [70], edge-magic labelings of unions of triangles were discussed. Denote the vertices of the i-th triangle as x_i, y_i, z_i and the edges as X_i, Y_i, Z_i.

Exercise 2.17 *Show that for a component i of an edge-magic sC_3, $X_i - x_i = Y_i - y_i = Z_i - z_i$. This value will be referred to as the common difference.*

The following two theorems give constructions for edge-magic labelings of sC_3 for s odd and even, respectively.

Theorem 2.31 *Let s be an odd positive integer. There exists an EMTL of sC_3 such that the labels $1, 2, \ldots, s$ occur on the vertices of different components and such that the common difference of every component is $3s$. The magic sum of this EMTL is $k = \frac{1}{2}(21s + 3)$.*

Proof. Since every common difference is $3s$, we need to only declare the set $\{x_i, y_i, z_i\}$ of edge labels for each component $i = 1, 2, \ldots, s$. These triples must partition the set $\{1, 2, \ldots, 3s\}$ such that the sum of the elements in any triple is constant. An explicit formula is found by letting $s = 2q - 1$ and $x_i = i$, $y_i = 4q - 2i$ if $1 \leq q$ and $y_i = 6q - 1 - 2i$ if $q + 1 \leq i \leq s$. Let $z_i = 5q - 3 + i$ if $1 \leq i \leq q$ and $z_i = 3q - 2 + i$ if $q + 1 \leq i \leq s$. This can easily be verified to be edge-magic. □

Theorem 2.32 *Let $s = 2t$ be a positive even integer. If $t \geq 3$ there exists an EMTL of sC_3 such that the labels $1, 2, \ldots, s - 2$ occur on the vertices of different components and such that the common difference of each of these components is $3s$. The magic sum is $k = 21t - 1$.*

Proof. The two components without a common difference of $3s$ are labeled edge-vertex-edge-vertex-edge-vertex as follows: $[2t, 9t, 9t - 1, 3t, 8t, 10t - 1]$ and $[3t - 1, 6t, 11t, 4t - 1, 5t, 12t]$. For the remaining components, we will list only the set of edges for each component. This is sufficient as every common difference is $d = 3s$. We distinguish between two cases:

Case 1: t is odd. Let $t = 2r + 1$. The edge sets are:

$$\begin{aligned}
&\{i, 5t + i, 4t - 1 - 2i\}, &&\text{for } 1 \leq i \leq r - 1 \\
&\{r, 4t, 4t + r\} \quad \text{and} &&\{r + 1, 4t - 2, 4t + r + 1\}, \\
&\{i, 5t - 2 + i, 4t + 1 - 2i\}, &&\text{for } r + 2 \leq i \leq t + 1 \\
&\{i, 3t - 1 + i, 6t - 2i\}, &&\text{for } t + 2 \leq i \leq t + r \\
&\{i, 3t + 1 + i, 6t - 2 - 2i\}, &&\text{for } t + r + 1 \leq i \leq 2t - 2.
\end{aligned}$$

Case 2: t is even. Let $t = 2r$. The edge sets are:

$$\begin{aligned}
&\{i, 5t + i, 4t - 1 - 2i\}, &&\text{for } 1 \leq i \leq r - 1 \\
&\{r, 4t, 4t + r - 1\} \quad \text{and} &&\{r + 1, 4t - 2, 4t + r\}, \\
&\{i, 5t - 2 + i, 4t + 1 - 2i\}, &&\text{for } r + 2 \leq i \leq t + 1 \\
&\{i, 3t - 1 + i, 6t - 2i\}, &&\text{for } t + 2 \leq i \leq t + r - 1 \\
&\{i, 3t + 1 + i, 6t - 2 - 2i\}, &&\text{for } t + r \leq i \leq 2t - 2.
\end{aligned}$$

Since the common difference is $3s$, the edge-magic property can be verified by checking that the sum of the edge labels on each component is $9t - 1$. □

Exercise 2.18 *Show that $2C_3$ does not have an edge-magic labeling.*

The following lemma can be used to give other possible values for the magic sum.

Lemma 2.33 *Let s be a positive integer and let λ_0 be an EMTL of sC_3 with magic sum of k. Assume that one of the components has edge labeled 1 and a common difference of $3s$. Then there exists another EMTL λ_1 of sC_3 with a magic sum of $k - 3$.*

Proof. Let T be a component of sC_3 as in the statement of the lemma. Thus, λ_0 labels T as $[1, x + 3s, y, 1 + 3s, x, y + 3s]$ for some x, y. Define $\lambda_1(z) = \lambda_0 - 1$ for any vertex or edge z which is not in T. Clearly $wt_{\lambda_1(v)} = k - 3$ for any vertex v not in T. Let λ_1 label T as $[6s, x - 1, y + 3s - 1, 3s, x + 3s - 1, y - 1]$. The fact that λ_1 is a total labeling is easily checked, noting that the label 1 was replaced by $6s$. Also, one can easily check the magic property holds. \square

Using Lemma 2.33, other values for the magic sum can be found for sC_3 when s is both even and odd. In fact, when s is even, s other possible magic sums are found and when s is odd, $s - 2$ other magic sums are found. In [70], the exact spectrum of sC_3 is found when $s = 2 \cdot 3^k$ with $k \geq 1$. Small cases of sC_3 have been checked by computer, and it was shown that $3C_3$ does not have an EMTL with magic constant 25, 26, 31, or 32. Furthermore, there is no EMTL for $4C_3$ with a magic constant of 32 or 43. But, an EMTL was found for each of the feasible values for $5C_3$. The spectrum of sC_3 has not been studied for $s \geq 6$, but the following conjecture was made.

Research Problem 2.16 *Prove or disprove the conjecture that for each $s \geq 5$, the spectrum of sC_3 coincides with its range of feasible values.*

Research Problem 2.17 *Prove or disprove that nK_4 is edge-magic when n is even.*

2.8.3 Unions of Stars

Suppose G is the union of two stars, $G = K_{1,m} \cup K_{1,n}$. Then $v(G) = m + n + 2$ and $e(g) = m + n$. If m and n are both odd, then all the vertices have odd degree, $e(G)$ is even and $e(G) + v(G) = 2m + 2n + 2 \equiv 2 \pmod{4}$, so by Corollary 2.3.1 G is not edge-magic. So let us assume n is even, say $n = 2t$.

Theorem 2.34 [82] *The graph $K_{1,m} \cup K_{1,2t}$ is edge-magic for any positive integers m and t.*

Proof. Denote the vertices of $K_{1,m}$ by x_0 (center), x_1, x_2, \ldots, x_m, and those of $K_{1,n}$ by y_0 (center), y_1, y_2, \ldots, y_{2t}. Then define a function λ by

$$\begin{aligned}
\lambda(x_0) &= 2 + 2m + 3t, \\
\lambda(x_i) &= i && \text{for } 1 \le i \le m, \\
\lambda(y_0) &= 1 + m + t, \\
\lambda(y_j) &= m + j && \text{for } 1 \le i \le t, \\
\lambda(y_j) &= 1 + m + j && \text{for } t + 1 \le i \le 2t, \\
\lambda(x_0 x_i) &= 2 + 2m + 2t - i && \text{for } 1 \le i \le m, \\
\lambda(y_0 y_j) &= 3 + 2m + 4t - j && \text{for } 1 \le j \le t, \\
\lambda(y_0 y_j) &= 2 + 2m + 4t - j && \text{for } t + 1 \le j \le 2t.
\end{aligned}$$

This is seen to be an edge-magic labeling with sum $4 + 4m + 5t$. □

(This construction is taken from [47].)

In fact, it is shown in [47] that $K_{1,m} \cup K_{1,n}$ is super-edge magic if and only if either m is a multiple of $n + 1$ or n is a multiple of $m + 1$.

It follows from Theorem 2.34 that the union $K_{1,1} \cup K_{1,2}$ of the two trivial stars P_2 and P_3 is edge-magic. We also know that the union of m copies of P_2 is edge-magic if and only if m is odd (see Exercise 2.15). More generally, we have

Theorem 2.35 [47] *Provided $n \ge 1$, $mP_2 + nP_3$ is always edge-magic.*

The proof consists of exhibiting labelings in several cases. It is further shown in [47] that all these graphs are super edge-magic, except for $2P_3$.

Exercise 2.19 *Verify that $2P_3$ has no super-magic edge labeling.*

2.8.4 Trichromatic Graphs

Graph colorings were defined and discussed briefly in Chap. 1. Formally, suppose $C = \{c_1, c_2, \dots\}$ is a set of undefined objects called *colors*. A *C-coloring* (or *C*-vertex coloring) ξ of a graph G is a map

$$\xi : V(G) \to C.$$

The sets $V_i = \{x : \xi(x) = c_i\}$ are called *color classes*. A *proper coloring* of G is a coloring in which no two adjacent vertices belong to the same color class. In other words,

$$x \sim y \Rightarrow \xi(x) \ne \xi(y).$$

A proper coloring is called an *n-coloring* if C has n elements. If G has an n-coloring, then G is called *n-colorable*. 2-colorable and 3-colorable graphs are also called *bipartite* and *trichromatic*, respectively.

A *total coloring* is an assignment ξ of colors to the vertices and edges of a graph G. A total coloring is called *proper* if the colors on a vertex and all edges that touch it contain no repeats. We shall refer to a graph with a proper total coloring in three colors as *totally trichromatic*. Ivančo and Lučkaničová [47] subsequently redefined total 3-colorings, calling them *e-m-colorings*.

Lemma 2.36 *Every 3-colorable graph is totally trichromatic.*

Proof. Suppose G is a 3-colorable graph. Select a proper 3-coloring $\xi : V(G) \to \{1, 2, 3\}$. Then define a 3-total coloring $\eta : V \cup E \to \{1, 2, 3\}$ by

$$\text{if } x \in V, \eta(x) = \xi(x)$$

$$\text{if } x \sim y, \{\eta(x), \eta(y), \eta(xy)\} = \{1, 2, 3\}. \qquad \square$$

Theorem 2.37 [52] *Say G is a 3-colorable edge-magic graph and H is the union of t disjoint copies of G, t odd. Then H is edge-magic.*

Proof. Suppose G has a proper total 3-coloring η, as guaranteed by Lemma 2.36, and suppose λ is an edge-magic total labeling of G with magic sum k. Denote the copies of G by G_0, G_1, \ldots, G_{2r}, where $t = 2r + 1$, and write s for $v + e$. Write $A = a_{ij}$ for the matrix (1.2) that was constructed in Sect. 1.1.3. Then vertex x of G_i receives label

$$\lambda(x) + sa_{\eta(x),i}$$

and edge xy of G_i receives label

$$\lambda(xy) + sa_{\eta(xy),i}.$$

This is an edge-magic total labeling with magic sum $3sr + k$. $\qquad \square$

The trichromatic graphs include all cycles and paths, so tC_v and tP_v are edge-magic for all odd t. (This was later independently proven by Wijaya and Baskoro [99], who gave a direct construction.)

The wheel W_n is trichromatic for n even. The union of an odd number of W_n's is not edge-magic when $n \equiv 3 \pmod 4$ (see Exercise 2.2), but the case $n \equiv 1 \pmod 4$ remains in doubt.

Research Problem 2.18 *Is tW_n edge-magic when $n \equiv 1 \pmod 4$ and t is odd?*

The single edge K_2 is of course trichromatic, so half of Exercise 2.15 follows from Theorem 2.37.

2.9 Super Edge-Magic Total Labelings

Recall that a $(V\text{-})$super edge-magic total labeling is one in which the vertex-labels are the integers $1, 2, \ldots, v$. As we noticed in Sect. 2.1.2, equation (2.3) can sometimes be used to show that a graph is not super edge-magic: if vertex x_i has degree d_i and is to receive label a_i, it is necessary to find an arrangement $\{a_i\}$ of the first v integers that makes $\sigma_0^{v+e} + \sum(d_i - 1)a_i$ divisible by e. We used this to show that the even cycles are not super edge-magic. However, Theorem 2.18 provides a super edge-magic labeling of every odd cycle. We have already observed (in Sect. 2.4.2) that all paths are super edge-magic.

Exercise 2.20 *Prove that all edge-magic one-factors are super edge-magic.*

Suppose G has a super edge-magic total labeling λ. Excluding the trivial case where G has no edges, some edge receives label $v + e$, so the weight of that edge is at least $1 + 2 + (v + e)$. On the other hand, the edge with label $v + 1$ will have weight at most $(v - 1) + v + (v + 1)$. So

$$1 + 2 + (v + e) \le k \le (v - 1) + v + (v + 1). \tag{2.21}$$

So we have

Lemma 2.38 [25] *Any super edge-magic graph other than K_1 satisfies*

$$e \le 2v - 3. \tag{2.22}$$

This simple lemma has a number of consequences. For example, no wheel is super edge-magic, nor is any complete graph with more than three vertices.

Lemma 2.38 also implies that no regular graph of degree greater than 3 can be super edge-magic. Suppose G is a regular graph of degree 3 with an edge-magic labeling λ. Necessarily v is even, say $v = 2n$. Then $e = 3n$. From equation (2.5), $6kn = 4s + 5n(5n + 1)$ where s is the sum of the vertex labels. If the labeling is super, $s = n(2n + 1)$, so $6k = 4(2n + 1) + 5(5n + 1) = 33n + 9$, so n must be odd. The first case is $n = 3, v = 6$.

There are two cubic graphs on six vertices. One is the triangular prism, which is super edge-magic (one triangle receives labels $1, 2, 3$, and the other receives labels $5, 6, 4$ in the corresponding places). The other, $K_{3,3}$, has no super edge-magic total labeling, as was seen in the complete search in Sect. 2.5.1. In fact, this is an instance of a stronger result:

Theorem 2.39 [25] *The complete bipartite graph $K_{m,n}$ is super edge-magic if and only if $m = 1$ or $n = 1$.*

Proof. One of the labelings in Theorem 2.27 is super, so the "if" part is easy. Now assume $m \geq n > 1$. From Lemma 2.38, a super edge-magic $K_{m,n}$ must satisfy $mn \leq 2(m+n) - 3$, or $(m-2)(n-2) \leq 1$. So we only need to investigate cases $m = n = 3$ and $m = 2$.

We know from the complete enumerations that $K_{2,2}$ and $K_{3,3}$ are not super edge-magic. So assume $m = 2, n \geq 3$. (2.21) gives $3n + 5 \leq k \leq 3n + 6$, and these two cases are duals of each other. So let us assume that $K_{2,n}$ has a super edge-magic total labeling with $k = 3n + 6$. Denote the two vertex sets as U and W.

The largest (edge) label is $3n + 2$, so $x_1 x_2$ cannot be an edge. Say x_1 and x_2 belong to U. To accommodate the edge with label $3n + 1$, $x_1 x_3$ must be an edge, so $x_3 \in W$. Then $x_2 x_3$ is the edge with label $3n + 1$, so $x_1 x_4$ is not an edge, and $x_4 \in U$. At this stage we see that the only possible edge with label $3n$ is $x_1 x_5$. But then $x_5 \in W$, and the $K_{2,n}$ contains two edges with label $3n - 1$, namely $x_2 x_5$ and $x_3 x_4$. □

Exercise 2.21 *Prove that a graph G is super edge-magic if and only if there is a map λ from $V(G)$ onto $\{1, 2, \ldots, v\}$ such that*

$$\{\lambda(x) + \lambda(y) \mid xy \in E(G)\} \tag{2.23}$$

is a set of consecutive integers. [29]

Research Problem 2.19 *Which unions of disjoint cycles are super edge-magic?*

Exercise 2.22 *Suppose G is a bipartite graph with vertex-sets V_1 and V_2, of sizes v_1 and v_2, respectively. An edge-magic total labeling of G is super-strong if the elements of V_1 receive labels $\{1, 2, \ldots, v_1\}$ and the elements of V_2 receive $\{v_1 + 1, v_1 + 2, \ldots, v\}$. Prove that every super-strongly edge-magic bipartite graph satisfies $e \leq v - 1$.* [78]

Recall that an $(n, 2)$-kite consists of a cycle of length n with a 2-edge path attached to one vertex. Consider such a kite with vertices $x_1, x_2, \ldots, x_n, y, z$, where (x_1, x_2, \ldots, x_n) is the cycle and x_n, y, z is the tail.

Theorem 2.40 [79] *An $(n, 2)$-kite has a super edge-magic labeling if and only if n is even.*

Proof. Suppose λ is such a labeling, with magic sum k. If the sum of the vertex labels is V and the sum of the edge labels is E, then

$$V = \sum_{i=1}^{n+2} i = \tfrac{1}{2}(n+2)(n+3),$$

$$E = \sum_{i=n+3}^{2n+4} i = \tfrac{1}{2}(2n+4)(2n+5) - V$$

and the sum of the weights of all the edges is

$$k(n+2) = 2V + E + \lambda(x_n) - \lambda(z)$$
$$= \tfrac{1}{2}(n+2)(n+3) + \tfrac{1}{2}(2n+4)(2n+5) + \lambda(x_n) - \lambda(z).$$

So

$$k = \frac{n+3}{2} + (n+2)(2n+5) + \frac{\lambda(x_n) - \lambda(z)}{n+2}$$
$$= \text{an integer} + \frac{n+3}{2} + \frac{\lambda(x_n) - \lambda(z)}{n+2}.$$

Now $\lambda(x_n) - \lambda(z)$ is nonzero and less than $n+2$, and k is an integer. The only possibility is that $\frac{\lambda(x_n)-\lambda(z)}{n+2} = \tfrac{1}{2}$, and that n is even.

Now assume n is even; put $n = 2m + 2$, where m is a nonnegative integer. If m is odd, define a labeling λ with vertex labels

$$\lambda(y) = (3m+7)/2, \ \lambda(z) = (3m+5)/2, \ \lambda(x_n) = (m+1)/2,$$

and

$$\lambda(x_i) = \begin{cases} (m-i+1)/2 & \text{for } i = 2, 4, \ldots, m-1 \\ (3m-i+4)/2 & \text{for } i = 1, 3, \ldots, m \\ (3m-i+3)/2 & \text{for } i = m+1, m+3, \ldots, 2m \\ (3m-i+10)/2 & \text{for } i = m+2, m+4, \ldots, 2m+1, \end{cases}$$

and edge labels

$$\lambda(x_n y) = 3m+7, \ \lambda(yz) = 2m+5, \ \lambda(x_1 x_n) = 3m+6,$$
$$\lambda(x_m x_{m+1}) = 3m+8, \ \lambda(x_n x_1) = 3m+9,$$

and

$$\lambda(x_i x_{i+1}) = \begin{cases} 3m+i+9 & \text{for } 1 \le i \le m-1 \\ m+i+5 & \text{for } m+1 \le i \le 2m. \end{cases}$$

For odd m, use

$$\lambda(y) = (3m+8)/2, \ \lambda(z) = (3m+4)/2, \ \lambda(x_n) = m/2,$$
$$\lambda(x_{m+1}) = m, \ \lambda(x_{n-1}) = (3m+6)/2,$$

and

$$\lambda(x_i) = \begin{cases} (m+i)/2 & \text{for } i = 2, 4, \dots, m-2 \\ (3m+i+9)/2 & \text{for } i = 1, 3, \dots, m-1 \\ (m+i+2)/2 & \text{for } i = m, m+2, \dots, 2m \\ (i-m-10)/2 & \text{for } i = m+3, m+5, \dots, 2m-1, \end{cases}$$

and edge labels

$$\lambda(x_n y) = 3m+7, \quad \lambda(yz) = 2m+5, \quad \lambda(x_1 x_n) = 3m+8,$$
$$\lambda(x_{m-1}x_m) = 2m+6, \quad \lambda(x_n x_1) = 3m+6, \quad \lambda(x_{n-2}x_{n-1}) = 2m+7,$$

and

$$\lambda(x_i x_{i+1}) = \begin{cases} 3m+i+6 & \text{for } 1 \le i \le m-2 \\ 4m-i+10 & \text{for } i = m, m+1 \\ 5m-i+10 & \text{for } m+2 \le i \le 2m. \end{cases}$$

In both cases λ is a super edge-magic labeling with sum $5m+11$. □

Finally, why have we restricted ourselves to the case of V-super edge-magic labelings, and ignored the existence of E-super edge-magic labelings? The answer is that the two problems are equivalent. Suppose λ is a V-super edge-magic labeling, with magic sum k, of a graph G with v vertices and e edges. For each vertex x, define $\mu(x) = \lambda(x) + e$, and for each edge xy, define $\mu(xy) = \lambda(xy) - v$. Then

$$\mu(x) + \mu(y) + \mu(xy) = \lambda(x) + \lambda(y) + \lambda(xy) + 2e - v,$$

and μ is an E-super edge-magic labeling of G with magic sum $k + 2e - v$. The two problems are not the same for vertex-magic, however; see Sect. 3.10.

2.10 A Cycle with a Chord

The graph constructed by adding a chord to a cycle always satisfies Lemma 2.38. We could ask wonder whether all such graphs are super edge-magic. The short answer is "no"—an exception is the graph formed from a six-cycle by joining two vertices of distance 2. But maybe this is the only exception. We shall provide solutions for all cases of an odd cycle with a chord, and the majority of cases of an even cycle with a chord. This work comes from [64].

We shall write C_v^t to mean the graph constructed from a C_v by joining two vertices whose distance in the cycle is t. In this notation, we have just remarked that C_6^2 is not super edge-magic.

Suppose C_v^t has a super edge-magic total labeling, and suppose the endpoints of the chord receive labels a and b. Then (2.3) is

$$(v+1)k = 2(1+2+\dots+v) + a + b + ((v+1)+(v+2)+\dots+(2v+1)) \tag{2.24}$$

so $v + 1$ must divide $a + b + \delta(v + 1)$, where δ is 0 when v is even and $\frac{1}{2}$ when v is odd.

When v is even, say $v = 2n$, we have $a + b \equiv 0 \pmod{2n + 1}$, so $a + b = 0$ or $2n + 1$ or $4n + 2 \ldots$ and the condition $0 \le a, b \le 2n$ implies $a + b = 2n + 1$. From (2.24), $k = 2n + 1$ and the set of sums (2.23) is

$$\{n + 1, n + 2, \ldots, 2n + 1, \ldots, 3n + 1\}.$$

For odd $v = 2n + 1$ we get $a + b = n + 1$ or $3n + 3$ yielding sums

$$\{n + 1, n + 2, \ldots, 3n + 2\} \text{ or } \{n + 2, n + 2, \ldots, 3n + 3\},$$

respectively. In this case the vertex labels that induce a super edge-magic total labeling of any C_v^t will also induce such a labeling of C_v.

It may well happen that one vertex labeling of C_v provides super edge-magic total labelings of C_v^t for all possible chord lengths t. A labeling with this desirable property will be called *universal*.

2.10.1 Odd Cycles

A chord in C_v of length t is of course also a chord of length $v - t$. This means that we can restrict our attention to chords of length at most $\frac{1}{2}v$, or, in the case of odd cycles, to chords of odd length.

It is easy to see that the labeling of C_{2n+1} constructed in Theorem 2.18 provides chords of all lengths t where $t \equiv 3 \pmod 4$, so C_v^t is super edge-magic for every $t \equiv 3 \pmod 4$, when v is odd. However, we have constructed better labelings.

Theorem 2.41 *The following vertex labeling λ of C_{4m+3} is universal:*

i values	$\lambda(x_i)$
i even, $2 \le i \le 2m$	$2m + 3 + \frac{1}{2}i$
i even, $2m + 2 \le i \le 4m + 2$	$\frac{1}{2}(i + 2)$
i odd, $1 \le i \le 2m + 1$	$\frac{1}{2}(i + 1)$
i odd, $2m + 3 \le i \le 4m + 1$	$2m + \frac{1}{2}(i + 5)$
$i = 4m + 3$	$2m + 3$

Proof. It is easy to verify that λ yields edges with vertex-weights $\{2m + 3, 2m + 4, \ldots, 6m + 5\}$. So it induces a super edge-magic labeling of C_{4m+3}. To prove that it is a super edge-magic labeling of C_{4m+3}^t it suffices to exhibit a chord of length t with vertex-weight $2m + 2$ or $6m + 6$.

For $1 \leq i \leq m$, $\lambda(x_{2i-1}) = i$ and $\lambda(x_{4m+2-2i}) = 2m + 2 - i$. So the chord $x_{2i-1}x_{4m+2-2i}$ of length $4(m - i) + 3$ has vertex-weight $2m + 2$. This covers chords of lengths $\{3, 7, \ldots, 4m - 1\}$.

For $1 \leq i \leq m - 1$, $\lambda(x_{2i}) = 2m + 3 + i$ and $\lambda(x_{4m+1-2i}) = 4m + 3 - i$. So the chord $x_{2i}x_{4m+1-2i}$ of length $4(m - i) + 1$ has vertex-weight $6m + 6$. This covers chords of lengths $\{5, 9, \ldots, 4m - 3\}$.

Finally, $x_{4m+1}x_{4m+3}$ is a chord of length $4m + 1$ whose vertex-weight is $\lambda(x_{4m+1}) + \lambda(x_{4m+3}) = (4m + 3) + (2m + 3) = 6m + 6$. So the construction covers all lengths. □

No one has yet found a universal labeling for all cycles C_{4m+1}. We present two theorems, one of which has chords with the appropriate weight of all odd lengths except 5 and 9 (and, consequently, all even lengths except $4m - 4$ and $4m - 8$), and another for chords of all lengths congruent to 1 (or 0) modulo 4 other than $4m - 3$ (or 4). So all odd cycles with chords are super edge-magic.

It should be noted that the constructions fail when $m < 3$. However, universal labelings of C_5 and C_9 exist; see Fig. 2.10.

Fig. 2.10. Universal labelings of C_5 and C_9

Theorem 2.42 *The following vertex labeling λ induces a super edge-magic labeling of C_{4m+1}^t for every t except $t = 5, 9, 4m - 4, 4m - 8$, other than $m \geq 3$:*

i values	$\lambda(x_i)$
i even, $2 \leq i \leq 2m - 2$	$2m + 4 + \frac{1}{2}i$
i even, $2m \leq i \leq 4m$	$\frac{1}{2}i + 2$
i odd, $1 \leq i \leq 2m + 1$	$\frac{1}{2}(i + 1)$
i odd, $2m + 3 \leq i \leq 4m - 3$	$2m + \frac{1}{2}(i + 5)$
$i = 4m \pm 1$	$\frac{1}{2}(i + 7)$

Proof. Again it is easy to verify that λ induces a super edge-magic labeling of C_{4m+1}. To prove that it is a super edge-magic labeling of C_{4m+1}^t, it suffices to exhibit a chord of length t with vertex-weight $2m + 1$ or $6m + 3$.

For $1 \leq i \leq m - 1$, $\lambda(x_{2i-1}) = i$ and $\lambda(x_{4m-2-2i}) = 2m + 2 - i$. So the chord $x_{2i-1}x_{4m-2-2i}$ of length $4(m - i) - 1$ has vertex-weight $2m + 1$. This covers chords of lengths $\{3, 7, \ldots, 4n - 5\}$.

For $1 \leq i \leq m - 5$, $\lambda(x_{2m+2i+1}) = 3m + 3 + i$ and $\lambda(x_{2m-8-2i}) = 3m - i$. So the chord $x_{2m+2i+1}x_{2m-8-2i}$ of length $4i + 9$ has vertex-weight $6m + 3$. This covers chords of lengths $\{13, 17, \ldots, 4m - 11\}$.

Finally, we have the following chords: $x_{2m-1}x_{2m+1}$, length $4m - 1$, vertex-weight $m + (m + 1) = 2m + 1$; $x_{4m-7}x_{4m+1}$, length $4m - 7$, vertex-weight $(2m+4) + (4m - 1) = 6m + 3$; and $x_{4m-1}x_{4m-5}$, length $4m - 3$, vertex-weight $(2m + 3) + 4m = 6m + 3$.

So the construction covers all required lengths. $\qquad\square$

Notice that the construction fails when $m < 3$. When $m = 1$ two values are assigned to $\lambda(x_3)$ and $\lambda(x_5) > 5$; when $m = 2$ the value assigned to $\lambda(x_{4m-5})$ comes from the third line of the table, not the fourth line, so chord length $4m - 3$ does not arise.

Theorem 2.43 *The following vertex labeling λ induces a super edge-magic labeling of C_{4m+1}^t for every $t \equiv 1 \pmod 4$ except $4m - 3$:*

i values	$\lambda(x_i)$
i even, $2 \leq i \leq 2m - 2$	$2m + 3 + \frac{1}{2}i$
$i = 2m$	$m + 2$
i even, $2m + 2 \leq i \leq 4m - 2$	$2m + 2 + \frac{1}{2}i$
$i = 4m$	$2m + 2$
i odd, $1 \leq i \leq 2m + 1$	$\frac{1}{2}(i + 1)$
i odd, $2m + 3 \leq i \leq 4m - 1$	$\frac{1}{2}(i + 3)$
$i = 4m + 1$	$\frac{1}{2}(i + 5)$

Proof. It is easy to check that λ is an edge-magic labeling with vertex labels $\{1, 2, \ldots, 4m + 1\}$ whose vertex-weights are $\{2m + 2, 2m + 3, \ldots, 6m + 2\}$. for odd i, $1 \leq i \leq 2m - 3$, the chord $x_i x_{4m-2-i}$ has vertex-weight $2m + 1$ and length $4m - 4 - 2i$, so the labeling induces a super edge-magic labeling of $C_{4m+1}^{4m-4-2i}$. As i ranges from 1 to $2m - 3$, this produces labelings for chords of lengths $5, 9, \ldots, 4m - 7$. $\qquad\square$

2.10.2 Even Cycles

Suppose there is a super edge-magic labeling of C_{2n}^t. Suppose the sums of adjacent labels are $k, k + 1, \ldots, k + 2n$, where the sum of the labels on the chord

is $k + i$. If we add the sums for all edges of the cycle, we add the label on each vertex twice, so

$$\sum_{j=0}^{2n} (k + j) = k + i + 2 \sum_{j=1}^{2n} j,$$

so

$$(2n + 1)k + \frac{2n(2n + 1)}{2} = 2n(2n + 1) + k + i,$$

$$i = 2n \left(k - \frac{2n + 1}{2} \right).$$

As $1 \leq i \leq 2n$, and as k is an integer, the only possibility is for $k - \frac{2n+1}{2}$ to equal $\frac{1}{2}$, so that $i = n$ and $k = n + 1$. If a suitable labeling of the vertices is found, we may take the chord to be any pair of vertices whose labels add to $2n + 1$.

For example, consider the C_8 whose vertices are labeled (sequentially) $1, 4, 3, 8, 2, 6, 7, 5$. A chord could be added joining the vertices labeled $1, 8$, giving a super edge-magic labeling of C_8^3, or $2, 7$, giving a super edge-magic labeling of C_8^2. This example does not provide a super edge-magic labeling of C_8^4, but such labelings are available (an example is $1, 5, 3, 2, 8, 4, 7, 6$).

A complete search shows that the only examples for C_4 and C_6 are $1, 3, 4, 2$ and $1, 3, 2, 6, 4, 5$. These give super edge-magic labelings of C_4^2 and C_6^3. There is no such labeling of C_6^2.

2.10.3 Some General Constructions

Theorem 2.44 *The following vertex labeling λ induces a super edge-magic labeling of C_{4m}^t for all $t \equiv 2 \pmod 4$:*

i values	$\lambda(x_i)$
i even, $2 \leq i \leq 2m - 2$	$2m + \frac{1}{2}(i + 2)$
i even, $2m \leq i \leq 4m$	$\frac{1}{2}(i + 2)$
i odd, $1 \leq i \leq 2m - 1$	$\frac{1}{2}(i + 1)$
i odd, $2m + 1 \leq i \leq 4m - 1$	$2m + \frac{1}{2}(i + 1)$

Proof. There is no problem in seeing that the vertex labels run from 1 to $4m$ and the vertex-weights run from $2m + 1$ to $6m + 1$, omitting $4m + 1$.

When i is odd and $1 \leq i \leq 2m - 1$, the chord $x_i x_{4m-i}$, of length $4m - 2$, has vertex-weight $4m + 1$. So the construction induces super edge-magic labelings of C_{4m}^t in all the required cases. □

This construction gives no other cases: the other chords with the appropriate vertex-weight duplicate these lengths.

Theorem 2.45 *The following vertex labeling λ induces a super edge-magic labeling of C_{4m+2}^t, $m > 1$, for all odd t other than 5, and for $t = 2, 6$:*

i values	$\lambda(x_i)$
$i = 2$	$2m + 4$
$i = 4$	$4m + 2$
$i = 2m$	$m + 2$
$i = 2m + 2$	$m + 3$
i even, $6 \leq i \leq 2m - 2$	$2m + 2 + \frac{1}{2}i$
i even, $2m + 4 \leq i \leq 4m + 2$	$2m + \frac{1}{2}i$
i odd, $1 \leq i \leq 2m + 1$	$\frac{1}{2}(i + 1)$
i odd, $2m + 3 \leq i \leq 4m + 1$	$\frac{1}{2}(i + 5)$

Proof. Again, the vertex labels have the required values, and the vertex-weights cover all possibilities once each, with $4m + 3$ missing. We see that the chords with vertex-weight $4m + 3$ are:

$x_{2j+1}x_{4m+4-2j}$, for $j = 1, 2, \ldots, m$, producing chords of all lengths $\equiv 3 \pmod 4$;

$x_{2m+2j+1}x_{4m+2j}$, for $j = 1, 2, \ldots, m - 5$, giving chords of all lengths $\equiv 1 \pmod 4$ other than 5;

$x_{2m-2}x_{2m}$ and $x_{2m-4}x_{2m+2}$, giving chords of lengths 2 and 6. □

Complete searches have been made for small cases, up to 12 vertices; see [64]. For example, there are 308 labelings for 11 vertices, covering all possible chords, and 503 for 12.

Research Problem 2.20 *For which values of t does C_{2n}^t have a super edge-magic labeling?*

Research Problem 2.21 *For which n is there a universal vertex labeling of C_{2n}?*

2.11 Edge-Magic Injections

Recall that an edge-magic injection is like an edge-magic total labeling, except that the labels can be any positive integers. We define an $[m]$-edge-magic injection of G to be an edge-magic injection of G in which the largest label is m, and call

m the *size* of the injection. The *edge deficiency* $\operatorname{def}_e(G)$ of G to be the minimum value of $m - v(G) - e(G)$, such that an $[m]$-edge-magic injection of G exists.

Theorem 2.46 *Every graph has an edge-magic injection.*

Proof. Suppose G is a graph with v vertices and e edges. The empty graph is trivially edge-magic, so we assume that G has at least one edge. Let(a_1, a_2, \ldots, a_v) be any Sidon sequence of length v with first element $a_1 = 1$. Define $k = a_{v-1} + 2a_v + 1$.

We now construct a labeling λ as follows. Select any edge of G and label its endpoints with a_{v-1} and a_v, and label the remaining vertices with the other members of the Sidon sequence in any order. If xy is any edge, define $\lambda(xy) = k - \lambda(x) - \lambda(y)$. Every edge weight will be equal to k. The smallest edge label will be $k - a_{v-1} - a_v = a_v + 1$, which is greater than any vertex label. If two edge labels were equal, say $\lambda(xy) = \lambda(zt)$, then $\lambda(x) + \lambda(y) = \lambda(z) + \lambda(t)$, and as the labels of vertices are members of a Sidon sequence this implies that $xy = zt$. The vertex labels are distinct by definition. So λ is an edge-magic injection. □

The proof of Theorem 2.46 gives us an upper bound on the deficiency:

Corollary 2.46.1 *If G is a graph with v vertices and (a_1, a_2, \ldots, a_v) is any Sidon sequence of length v with $a_1 = 1$, then*

$$\operatorname{def}_e(G) \leq a_{v-1} + 2a_v - a_2 - v - e(G).$$

Proof. In the above construction, no label can be greater than $k - 1 - a_2$. □

This upper bound will not usually be very good. For example, consider the graph constructed from C_5 by joining two inadjacent vertices. Using the Sidon sequence $(1, 2, 3, 5, 8)$, a labeling with $k = 22$ is obtained, and the best assignment of the sequence to the vertices gives largest label 17, and deficiency 6. However, the graph is actually edge-magic. See Fig. 2.11.

Exercise 2.23 *If G is an incomplete graph with v vertices and (a_1, a_2, \ldots, a_v) is any Sidon sequence of length v with $a_1 = 1$, prove that*

$$\operatorname{def}_e(G) < a_{v-1} + 2a_v - a_2 - v - e(G).$$

In the case of complete graphs, we have essentially encountered the edge-magic deficiency already:

Theorem 2.47 *The edge-magic deficiency of K_v equals the magic number $M(v)$.*

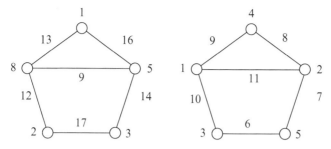

Fig. 2.11. Deficiency 6 on *left*; magic on *right*

Proof. Consider a edge-magic total labeling λ of $K_v \cup M(v)K_1$. This graph has $v + M(v)$ vertices and $e(K_v)$ edges, so the largest label is $v + M(v) + e(K_v)$, and clearly this label occurs on a vertex or edge of K_v. The labeling constructed by restricting λ to K_v is an $[v + M(v) + e(K_v)]$-edge-magic injection of K_v. Obviously any injection of size $v + m + e(K_v)$ gives rise to an edge-magic total labeling of $K_v \cup mK_1$ (apply the m unused labels to the extra vertices), so $v + M(v) + e(K_v)$ is the smallest possible size, and $def_e(K_v) = M(v)$. ☐

From Theorem 2.47 it is clear that magic number and edge-magic deficiency are essentially equivalent, but the two concepts are rather different when applied to vertex-magic labelings.

Exercise 2.24 *Suppose G is a graph with v vertices. Prove that*

$$\mathrm{def}_e(G) \leq M(v) + \binom{v}{2} - e(G).$$

Both Corollary 2.46.1 and Exercise 2.24 give crude upper bounds for the $\mathrm{def}_e(G)$. Very little work has been done on finding good bounds for edge-magic deficiencies, and the only known families of graphs for which the exact values are known are the various families of edge-magic graphs (which of course have edge-magic deficiency 0). Further results on edge-magic injections can be found in [73].

3

Vertex-Magic Total Labelings

3.1 Basic Ideas

3.1.1 Definitions

A one-to-one map λ from $E(G) \cup V(G)$ onto the integers $\{1, 2, \ldots, e + v\}$ is a *vertex-magic total labeling* if there is a constant h so that for every vertex x,

$$\lambda(x) + \sum \lambda(xy) = h \tag{3.1}$$

where the sum is over all vertices y adjacent to x. So the magic requirement is $\text{wt}(x) = h$ for all x. The constant h is called the *magic constant* for λ. Again, a graph with a vertex-magic total labeling will be called *vertex-magic*.

It is not hard to find examples of vertex-magic total labelings for some graphs. One labeling for the graph $K_4 - e$ is shown in Fig. 3.1. On the other hand, not every graph has a vertex-magic labeling. For the graph K_2, $\lambda(x) \neq \lambda(y)$ implies $\lambda(x) + \lambda(xy) \neq \lambda(y) + \lambda(xy)$, so no labeling is possible. Similarly, any isolated vertex x must have $\lambda(x) = h$, so the prohibition of repeated labels means that there can be at most one isolate. These easy observations will be important, so we state them as a theorem:

Theorem 3.1 *If G has an isolated edge or two isolated vertices, then G is not vertex-magic.*

A.M. Marr and W.D. Wallis, *Magic Graphs*, DOI 10.1007/978-0-8176-8391-7_3, © Springer Science+Business Media New York 2013

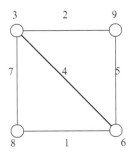

Fig. 3.1. A vertex-magic total labeling of $K_4 - e$

Exercise 3.1 *Suppose we wish to define a labeling that is vertex-magic, and uses the labels $\{0, 1, \ldots, v + e - 1\}$ once each. Prove that in such a labeling, 0 cannot be the label of any vertex of degree 1. Prove that P_3 has no such labeling, but P_4 has one.*

3.1.2 Basic Counting

Let s_v denote the sum of the vertex labels and s_e the sum of the edge labels in a vertex-magic total labeling λ. Clearly, since the labels are the numbers $1, 2, \ldots, v + e$, the sum of all labels is

$$s_v + s_e = \sigma_0^{v+e} = \binom{v+e+1}{2}.$$

At each vertex x_i we have $\lambda(x_i) + \sum \lambda(x_i y) = h$. Summing this over all v vertices x_i is equivalent to adding each vertex label once and each edge label twice, so

$$s_v + 2s_e = vh. \tag{3.2}$$

Combining these two equations gives

$$s_e + \binom{v+e+1}{2} = vh. \tag{3.3}$$

The edge labels are all distinct (as are all the vertex labels). The edges could conceivably receive the e smallest labels or, at the other extreme, the e largest labels, or anything between. Consequently we have

$$\sigma_0^e \le s_e \le \sigma_v^{v+e} \tag{3.4}$$

A similar result holds for s_v. Combining (3.3) and (3.4), we get

$$\binom{v+e+1}{2} + \binom{e+1}{2} \le vh \le 2\binom{v+e+1}{2} - \binom{v+1}{2} \tag{3.5}$$

which gives the range of feasible values for h.

It is clear from (3.1) that when h is specified and the edge labels are known, then the vertex labels are determined. So the labeling is completely described by the edge labels. Surprisingly, however, the vertex labels do not completely determine the labeling. Having assigned the vertex labels to a graph, it may be possible to assign the edge labels to the graph in several different ways. Figure 3.2 shows two vertex-magic total labelings of W_4 that have the same vertex labeling but different edge labelings.

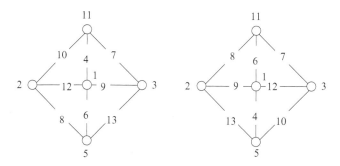

Fig. 3.2. Vertex-magic total labelings of W_4 with the same vertex-labels

Research Problem 3.1 *Construct an infinite family of graphs, each having two vertex-magic total labelings with the same vertex labeling but different edge labelings.*

3.2 Regular Graphs

If a *regular* graph possesses a vertex-magic total labeling, we can create a new vertex-magic total labeling from it. Given the vertex-magic total labeling λ for graph G, define the map λ' on $E(G) \cup V(G)$ by

$$\lambda'(x) = v + e + 1 - \lambda(x)$$

for any vertex x, and

$$\lambda'(xy) = v + e + 1 - \lambda(xy)$$

for any edge xy. Clearly λ' is also a one-to-one map from the set $E(G) \cup V(G)$ to $\{1, 2, \ldots, e + v\}$. Just as in the case of edge-magic total labelings, we shall call λ' the *dual* of λ. In contrast to the edge-magic case, we have the following theorem:

Theorem 3.2 *The dual of a vertex-magic total labeling for a graph G is vertex-magic if and only if G is regular.*

Proof. Suppose λ is a vertex-magic total labeling for G in which $\text{wt}(x) = h$. Then the dual λ' satisfies

$$wt_{\lambda'}(x) = \lambda'(x) + \sum \lambda'(xy)$$
$$= v + e + 1 - \lambda(x) + \sum (v + e + 1 - \lambda(xy))$$
$$= (r+1)(v+e+1) - h$$

where r is the number of edges incident at x. Clearly this is constant if and only if r is constant; in other words, if and only if G is regular. Then $(r+1)(v+e+1) - h$ is the magic constant for the dual. □

The general problem of whether one can use a vertex-magic total labeling of a graph G to produce such a labeling of some subgraph or supergraph of G appears to be very difficult. The next theorem answers a very special case of this question for regular graphs.

Theorem 3.3 *Let G be a regular graph having a vertex-magic total labeling in which the label 1 is assigned to some edge e'. Then the graph $G - e'$ has a vertex-magic total labeling.*

Proof. Suppose G is an r-regular graph. Define a new mapping Λ by $\Lambda(x_i) = \lambda(x_i) - 1$ and $\Lambda(xy) = \lambda(xy) - 1$. Now Λ is a one-to-one mapping from $E \cup V$ to $\{0, 1, 2, \ldots, e + v - 1\}$. If we now delete the edge e' (which is labeled 0 by Λ) from G, then Λ is a one-to-one map from $(E - e') \cup V$ to $\{0, 1, 2, \ldots, e + v - 1\}$. Also the weight of x under Λ is

$$wt_{\Lambda}(x) = \lambda(x) - 1 + \sum (\lambda(xy) - 1)$$
$$= wt_{\lambda}(x) - r - 1$$
$$= h - r - 1.$$

Thus $wt_{\Lambda}(x)$ is a constant and Λ is a vertex-magic total labeling for $G - e'$. □

Theorem 3.4 *If G is a regular graph and e an edge such that $G - e$ has a vertex-magic total labeling, then that labeling is derived from a vertex-magic total labeling of G by the process described in Theorem 3.3.*

Proof. Suppose G is an r-regular graph and let λ be any labeling for $G - e$ where e is the edge xy. Adjoin the edge xy to $G - e$ and define $\lambda(xy) = 0$. Now adding 1 to all the labels defines a new mapping λ' that is easily seen to be a vertex-magic total labeling of G with label 1 on edge xy. If the magic constant of λ is h, then the magic constant of λ' is $h + r + 1$. □

James MacDougall [60] has conjectured that all regular graphs other than K_2 and $2K_3$ are vertex-magic. Up to now no counterexamples to MacDougall's

conjecture have been found. Gray [36] has added significant support to this conjecture by demonstrating that "almost all" regular graphs of odd order possess vertex-magic total labelings, as do many graphs of even order.

3.3 Cycles and Paths

The easiest regular graphs to deal with are the cycles. For cycles (and only for cycles) a vertex-magic total labeling is equivalent to an edge-magic total labeling, and the edge-magic labelings have already received some attention in Sect. 2.4.

Theorem 3.5 *The n-cycle C_n has a vertex-magic total labeling for any $n \geq 3$.*

Proof. Suppose C_n is the cycle (x_1, x_2, \ldots, x_n). The constructions of Theorems 2.18–2.21 provide an edge-magic total labeling λ' for C_n for every $n \geq 3$. Denote the magic sum of λ' by h. Then define a new mapping λ by $\lambda(x_i) = \lambda'(x_i x_{i+1})$ and $\lambda(x_i x_{i+1}) = \lambda'(x_{i+1})$, where the subscripts are integers modulo n. Clearly h is the weight at each vertex, and λ is a vertex-magic total labeling of G. □

Corollary 3.5.1 *P_n, the path with n vertices, is vertex-magic for any $n \geq 3$.*

Proof. For each $n \geq 3$ at least one of the edge-magic labelings λ' referred to in the proof of the theorem assigns the label 1 to a vertex. Then the corresponding vertex-magic labeling λ will assign the label 1 to some edge e. By Theorem 3.3, $C_n - e$ will have a vertex-magic labeling. □

Corollary 3.5.2 *Every vertex-magic total labeling of P_n is derived from a vertex-magic total labeling of C_n.*

Proof. This follows immediately from Theorem 3.4. □

Using equation 3.5, we can readily determine the feasible values of h for the n-cycle. We find

$$\frac{5v + 3}{2} \leq h \leq \frac{7v + 3}{2}.$$

A systematic search has found exactly 4 labelings for the 3-cycle, one for each feasible value of h. As mentioned above, once h is given, the edge labels completely determine the labeling, so we list only the edge labels.

$$
\begin{array}{rl}
h = & 9 \quad 1, 2, 3 \\
h = & 10 \quad 1, 3, 5 \\
h = & 11 \quad 2, 4, 6 \\
h = & 12 \quad 4, 5, 6
\end{array}
$$

Since cycles are regular graphs, the duality described in Sect. 3.2 applies. The labeling with $h = 12$ is dual to the labeling with $h = 9$, and the labeling with $h = 11$ is dual to that with $h = 10$.

There are exactly six vertex-magic total labelings for the 4-cycle and again they come in dual pairs. Once more every feasible value of h admits a labeling. They are listed below, with the edge labels given in cyclic order:

$$
\begin{array}{ll}
h = 12 & 1, 3, 2, 6 \\
h = 13 & 1, 4, 6, 5 \\
h = 13 & 1, 5, 2, 8 \\
h = 14 & 3, 4, 8, 5 \\
h = 14 & 1, 7, 4, 8 \\
h = 15 & 3, 7, 6, 8.
\end{array}
$$

For the 5-cycle, we find $14 \leq h \leq 19$. There are again 6 labelings. None exists corresponding to $h = 15$ or $h = 18$. For $h = 14$ there is a unique solution $(1, 4, 2, 5, 3)$, and for $h = 16$ we find $(1, 5, 9, 3, 7)$ and also $(1, 7, 3, 4, 10)$. Each of these has a dual.

The notion of vertex-magic labeling was at least partially suggested by the following question which appeared on a set of mathematical enrichment problems for high-school students [89] (The problem appeared at the time when corruption charges against International Olympic Committee members were dominating the news):

The Olympic emblem consists of five overlapping rings containing 9 regions. In order to contribute to a pension fund for a retiring IOC delegate, people are asked to deposit money into each region. The guidelines allow the delegate to take all the money in any one of the rings. Place $1, $2, \ldots, 9 in the nine regions so that the amount in each ring is the same.

This problem can be interpreted as asking for a labeling on the path of five vertices; one of the solutions the students found is shown in Fig. 3.3.

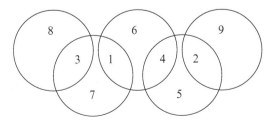

Fig. 3.3. Solution of the olympic rings problem

In the case of a path with v vertices, we have $e = v - 1$ and $v + e = 2v - 1$, and therefore

$$s_e + v(2v - 1) = vh.$$

So $s_e \equiv 0 \pmod{v}$ and consequently, from (3.2), $s_v \equiv 0 \pmod{v}$ also. Equations (3.3) and (3.4) then give

$$\frac{v(v-1)}{2} \le s_e \le 3\frac{v(v-1)}{2}$$

from which it follows that

$$\frac{5v - 3}{2} \le h \le \frac{7v - 5}{2}. \tag{3.6}$$

For a path with three vertices, (3.6) implies that $6 \le h \le 8$. According to Theorem 3.4, the vertex-magic total labelings of P_3 will be derived from those of C_3 that assign the label 1 to an edge. There are two such listed previously so there are two vertex-magic total labelings for P_3; they have sums 6 and 7. There is no vertex-magic total labeling of P_3 with magic constant $h = 8$.

For the path with four vertices, equation (3.6) implies $9 \le h \le 11$. Four of the six labelings of the 4-cycle listed above provide a derived labeling for the 4-path. This time there is a labeling for all feasible values of h (we list the edge labels only):

$$\begin{array}{rl}
h = & 9 \quad 2, 1, 5 \\
h = & 10 \quad 4, 5, 3 \\
h = & 10 \quad 4, 1, 7 \\
h = & 11 \quad 6, 3, 7.
\end{array}$$

For $v = 5$, equation (3.6) implies $11 \le h \le 15$. There are again four labelings; they correspond to $h = 11, 13$ and 14. There is no labeling with $h = 12$ or $h = 15$.

$$\begin{array}{ll}
h = 11 & 2, 4, 1, 3 \\
h = 12 & \text{none} \\
h = 13 & 4, 8, 2, 6 \\
h = 13 & 9, 3, 2, 6 \\
h = 14 & 6, 7, 3, 9 \\
h = 15 & \text{none.}
\end{array}$$

Further systematic search discovered ten labelings for $v = 6$; there are 66 for $v = 7$, and for $v = 8$ there are 131 labelings.

Exercise 3.2 *Prove that there is no vertex-magic total labeling of P_3 with magic constant $h = 8$.*

Exercise 3.3 *Find a vertex-magic total labeling of the Petersen graph.*

Exercise 3.4 *The triangular book $B_{3,n}$ was defined in Exercise 2.10. Prove that $B_{3,n}$ is vertex-magic if and only if $n \leq 3$.*

3.4 Vertex-Magic Total Labelings of Wheels

3.4.1 Large Wheels Cannot be Labeled

Suppose W_n is the wheel whose n rim vertices form the cycle (x_1, x_2, \ldots, x_n). W_n has $v = n + 1$ and $e = 2n$, so the labels are the numbers in the set $\{1, 2, \ldots, 3n + 1\}$. The inequalities (3.4) yield

$$\frac{13n^2 + 7n + 3}{2(n+1)} \leq h \leq \frac{17n^2 + 15n + 4}{2(n+1)} \tag{3.7}$$

but these inequalities become largely irrelevant when the structure of W_n is taken into account. In fact, the permissible values of h are determined by the degree of the hub vertex. If the degree of the hub vertex is too high, in other words n is too large, then no labeling is possible:

Theorem 3.6 *The wheel W_n has no vertex-magic total labeling when $n > 11$.*

Proof. Denote the hub vertex of W_n by u and let x_1, \ldots, x_n be the rim vertices. Then

$$\begin{aligned}
h &= \mathrm{wt}(u) \\
&\geq \sigma_0^{n+1} \\
&= \tfrac{1}{2}(n+1)(n+2). \tag{3.8}
\end{aligned}$$

Next we consider the sum of weights of all the rim vertices. An upper bound for this sum is found by assigning the n largest labels to the rim edges (because they are each counted twice), and the $2n$ next largest labels to the rim vertices and the spoke edges. We find

$$\begin{aligned}
\mathrm{wt}(x_1) + \cdots + \mathrm{wt}(x_n) &\leq \sigma_1^{2n+1} + 2\sigma_{2n+1}^{3n+1} \\
&= \sigma_1^{3n+1} + \sigma_{2n+1}^{3n+1} \\
&= n(7n+6).
\end{aligned}$$

Since there are n rim vertices,

$$h \leq 7n + 6. \tag{3.9}$$

It is easy to see that expression (3.9) is less than expression (3.8) for all $n > 11$. So no labeling is possible. $\qquad\square$

Exercise 3.5 *Prove that the* fan F_n *has no vertex-magic total labeling when* $n > 10$. [63]

Exercise 3.6 *The* friendship graph T_n *consists of* n *triangles with a common vertex. Prove that* T_n *is vertex-magic if and only if* $n \leq 3$. [63]

3.4.2 Small Wheels can have many Labelings

The labelings for W_3, W_4, and W_5 have been enumerated using a computer search. The feasible values of h are determined by whichever of (3.7) or (3.9) and (3.8) are the more restrictive $N_n(h)$ is the number of labelings of W_n with magic constant h.

h	$N_3(\mathbf{h})$
19	–
20	2
21	5
22	–
23	5
24	2
25	–

h	$N_4(\mathbf{h})$
26	89
27	149
28	522
29	376
30	573
31	211
32	131
33	29

h	$N_5(\mathbf{h})$
32	239
33	1242
34	2694
35	5180
36	7873
37	7173
38	4124
39	2511
40	776
41	80

3.4.3 Generalizations of Wheels

Suppose G is the graph derived from a wheel by duplicating the hub vertex one or more times. We call G a *t-fold wheel* if there are t hub vertices, each adjacent to all rim vertices, and not adjacent to each other. A twofold wheel is shown in (Figure 3.4):

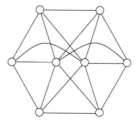

Fig. 3.4. A twofold wheel

A calculation like that of Theorem 3.6 obtains a similar result

Theorem 3.7 *Let G be a t-fold wheel with n rim vertices. There is a function $f(t)$ where, for $n > f(t)$, no labeling of G exists.*

Proof. In this case $v = n+t$ and $e = (t+1)n$, so $v+e = (t+2)n+t$. Assigning the smallest possible labels to the t hub vertices u_1, \ldots, u_t and the spoke edges, we find

$$\operatorname{wt}(u_1) + \cdots + \operatorname{wt}(u_t) \geq \sum_1^{tn+t} i$$
$$= \frac{(tn+t)(tn+t+1)}{2}$$

so that

$$h \geq \tfrac{1}{2}(n+1)(t(n+1)+1) \tag{3.10}$$

Assigning the n largest possible labels to the rim edges and the next $(t+1)n$ largest labels to the rim vertices and the spoke edges (i.e., all but the hub vertices), we have

$$\operatorname{wt}(v_1) + \cdots + \operatorname{wt}(v_n) \leq \sum_{t+1}^{M} i + \sum_{M-n+1}^{M} i$$
$$= 2\sigma_0^{v+e} - \sigma_0^t - \sigma_0^{v+e-n}$$
$$= \tfrac{1}{2}n\{n(t+6)(t+1) + (t+3)(2t+1)\}.$$

Since there are n rim vertices, we have

$$h \leq \frac{(t+6)(t+1)n + (t+3)(2t+1)}{2} \tag{3.11}$$

No labeling will be possible whenever the expression from (3.11) is less than that from equation (3.10), i.e. when the following inequality in n is satisfied:

$$tn^2 - (t^2 + 2t + 3)n - (2t^2 + 4t + 1) > 0$$

If we let $f(t)$ denote the larger root of this quadratic, the theorem follows. □

We note that the function $f(t)$ is asymptotically linear in t. When $t = 1$, we get the result from the previous theorem. For double wheels $(t = 2)$, there is no labeling for $n > 10$; for wheels with 3 hubs, no labeling exists for $n > 10$; and for wheels with 4 hubs, no labeling exists for $n > 11$.

3.5 Vertex-Magic Total Labelings of Complete Bipartite Graphs

We shall take the complete bipartite graph $K_{m,n}$ to have vertex-set

$$\{x_1, x_2, \ldots, x_m, y_1, y_2, \ldots, y_n\}$$

and edge-set

$$\{x_i y_j : 1 \leq i \leq m, 1 \leq j \leq n\}.$$

So a vertex-magic total labeling λ of $K_{m,n}$ can be represented by an $(m+1) \times (n+1)$ array

$$A = \begin{bmatrix} a_{00} & a_{01} & a_{02} & \cdots & a_{0n} \\ a_{10} & a_{11} & a_{12} & \cdots & a_{1n} \\ \cdot & \cdot & \cdot & \cdots & \cdot \\ \cdot & \cdot & \cdot & \cdots & \cdot \\ \cdot & \cdot & \cdot & \cdots & \cdot \\ a_{m0} & a_{m1} & a_{m2} & \cdots & a_{mn} \end{bmatrix} \qquad (3.12)$$

where

$$\begin{aligned} a_{00} &= 0 & a_{i0} &= \lambda(x_i) \\ a_{0j} &= \lambda(y_j) & a_{ij} &= \lambda(x_i y_j). \end{aligned}$$

The matrix A will be called the *representation matrix* of λ. The magic requirement is that all row-sums and column-sums, except for row 0 and column 0, must be equal (to h say), and that the $(m+1)(n+1)$ entries are $\{0, 1, \ldots, mn + m + n\}$ in some order.

We shall call a $K_{m,n}$ *unbalanced* if its parts differ in size by more than 1. We observe that an unbalanced $K_{m,n}$ cannot have a vertex-magic total labeling:

Theorem 3.8 [61] *If $K_{m,n}$ is unbalanced, then it has no vertex-magic total labeling.*

Proof. Without loss of generality, assume $m \leq n$. Suppose $K_{m,n}$ has a vertex-magic total labeling with magic constant h. For this graph $v = m + n$ and $e = mn$ so the label set is $\{1, 2, \ldots, mn + m + n\}$. The sum of the weights on $\{x_1, \ldots, x_m\}$ is at least the sum of all but the largest n labels, so

$$mh \geq \sigma_0^{mn+m}$$

$$= \frac{(mn+m)(mn+m+1)}{2};$$

$$h \geq \frac{(n+1)(mn+m+1)}{2} \qquad (3.13)$$

On the other hand, the sum of the weights on $\{y_1, \ldots, y_n\}$ is at most the total of all but the m smallest labels:

$$nh \leq \sigma_m^{mn+m+n}$$

$$= \frac{(mn+m+n)(mn+m+n+1) - m(m+1)}{2}$$

$$= \frac{(mn^2 + 2mn + n^2 + n)(m+1)}{2};$$

$$h \leq \frac{(mn + 2m + n + 1)(m+1)}{2}. \tag{3.14}$$

Combining (3.13) and (3.14),

$$(n+1)(mn+m+1) \leq (mn+2m+n+1)(m+1),$$

and on simplifying one obtains $m \geq n - 2 + \frac{2}{n+2}$, so $m \geq n - 1$. □

In particular, the only star that can have a vertex-magic total labeling is $K_{1,2}$.

3.5.1 Construction of Vertex-Magic Total Labelings of $K_{m,n}$

We now give constructions for vertex-magic total labelings of complete bipartite graphs in the cases not eliminated by Theorem 3.8.

Labeling $K_{m,m}$

Theorem 3.9 [61] *For every $m > 1$, $K_{m,m}$ has a vertex-magic total labeling with magic constant $\frac{1}{2}[(m+1)^3 - (m+1)]$.*

Proof. Let $S = (s_{ij})$ be any RCmagic square of order $m + 1$ on the numbers $\{1, \ldots, (m+1)^2\}$. (For convenience, assume that the rows and columns of S are numbered $0, 1, \ldots, m$.) Each row and column sums to the magic square constant $\frac{1}{2}(m+1)(m^2 + 2m + 2)$. Form the matrix $A = (a_{ij})$ where $a_{ij} = s_{ij} - 1$. Since S is magic, the rows and columns of A will each sum to the constant

$$h = \frac{1}{2}(m+1)(m^2 + 2m + 2) - (m+1) \tag{3.15}$$

and the entries of A will be the numbers in $\{0, \ldots, (m+1)^2 - 1\}$, once each. There are standard constructions ([2, 87]) for magic squares of all orders. We shall assume that the rows and columns of A are permuted so that $a_{00} = 0$. Then A is the representation matrix of a vertex-magic total labeling λ with h given by equation (3.15). The magic constant is easily checked. □

Labeling $K_{m,m+1}$, m Odd

A solution for $K_{1,2}$ is easily constructed. So let us write $m = 2n - 1$ where $n > 1$. The construction proceeds for a given n by defining two $(2n - 1) \times 2n$ matrices,

$A = (a_{ij})$ and $B = (b_{ij})$, and then using them to construct a $2n \times (2n + 1)$ representation matrix C. For consistency with the earlier notation, C has first row and column indexed with 0.

The value of a_{ij} depends on the parity of i and j, as well as their values. The formula is

$$a_{ij} = m + 1 - j \quad \text{if } i + j \text{ is odd, } i + j \leq m + 1$$
$$\text{or } i + j \text{ is even, } i + j > m + 1,$$
$$a_{ij} = j - 1 \quad \text{otherwise.}$$

We shall need the row and column sums of this matrix. If i is even, say $i = 2t$, the sum of elements in row i is

$$\sum_{j=1}^{2n} a_{ij} = \sum_{h=1}^{n} (a_{i,2h-1} + a_{i,2h})$$

$$= \sum_{h=1}^{n-t} (a_{i,2h-1} + a_{i,2h}) + \sum_{h=n-t+1}^{n} (a_{i,2h-1} + a_{i,2h})$$

$$= \sum_{h=1}^{n-t} ((2n - 2h + 1) + (2h - 1))$$

$$+ \sum_{h=n-t+1}^{n} ((2h - 2) + (2n - 2h))$$

$$= \sum_{h=1}^{n-t} 2n + \sum_{h=n-t+1}^{n} (2n - 2)$$

$$= 2n^2 - i.$$

If i is odd, say $i = 2t + 1$,

$$\sum_{j=1}^{2n} a_{ij}$$

$$= \sum_{h=1}^{n} (a_{i,2h-1} + a_{i,2h})$$

$$= \sum_{h=1}^{n-t} a_{i,2h-1} + \sum_{h=1}^{n-t-1} a_{i,2h} + \sum_{h=n-t+1}^{n} a_{i,2h-1} + \sum_{h=n-t}^{n} a_{i,2h}$$

$$= \sum_{h=1}^{n-t} (2h - 2) + \sum_{h=1}^{n-t-1} (2n - 2h)$$

$$+ \sum_{h=n-t+1}^{n} (2n - 2h + 1) + \sum_{h=n-t}^{n} (2h - 1)$$

$$= 2n^2 - i,$$

so in either case row i has sum $2n^2 - i$. The column sum is more easily calculated: column j contains n entries equal to $j - 1$ and $n - 1$ equal to $2n - j$. So for each j,

$$\sum_{i=1}^{2n-1} a_{ij} = n(j-1) + (n-1)(2n-j)$$

$$= 2n^2 - 3n + j.$$

Matrix B has first and last columns $2n - 2, 2n - 3, \ldots, 1, 0$:

$$b_{1j} = b_{mj} = m - i - 1.$$

The second column begins with the odd integers $2n - 5, 2n - 7, \ldots, 1$, then has the even integers $2n - 2, 2n - 4, \ldots, 0$ and ends with $2n - 3$. The other columns are formed by back-circulating this column. That is,

$$b_{i,j} = b_{i+1,j-1}$$

(with first subscript reduced mod $(2n)$ and second subscript reduced mod $(2n+1)$ where necessary).

In each row of B, columns 2 through $2n-1$ contain all the integers $0, 1, \ldots, m - 2$ except

$$x_i = 2n - 1 - 2i \text{ is missing from row } i, i = 1, \ldots, n - 1,$$
$$x_i = 4n - 2i - 2 \text{ is missing from row } i, i = n, n + 1, \ldots, 2n - 1.$$

So the row sums are

$$\sum_{j=1}^{m} b_{ij} = 2(2n - i - 1) + \sum_{h=0}^{2n-2} h - x_i$$

$$= 2n^2 + n - 1 - 2i - x_i$$

$$\sum_{j=1}^{m} b_{ij} = 2n^2 - n \text{ if } i \leq n - 1,$$

$$\sum_{j=1}^{m} b_{ij} = 2n^2 - 3n + 1 \text{ if } i \geq n.$$

Since each column is a permutation of $\{0, 1, \ldots, 2n - 2\}$, each column sum is

$$\sum_{i=1}^{2n-1} b_{ij} = 2n^2 - 3n + 1.$$

we now define a $2n \times (2n + 1)$ matrix C by

$$c_{00} = 0,$$
$$c_{0j} = 4n^2 + 2n - j, \ 1 \leq j \leq 2n,$$
$$c_{i0} = i, \ 1 \leq i \leq n - 1,$$
$$\phantom{c_{i0}} = 4n^2 - 2n + i, \ n \leq i \leq 2n - 1,$$
$$c_{ij} = a_{ij} + 2nb_{ij} + n, 1 \leq i \leq 2n - 1, 1 \leq j \leq 2n.$$

Theorem 3.10 *The matrix C is the representation matrix of a vertex-magic total labeling of $K_{2n-1,2n}$ with magic constant $4n^3 + 2n^2$.*

Proof. It is necessary to show that the sum of entries in every row and column of C (except possibly row 0 and column 0) equals $4n^3 + 2n^2$ and that every integer from 0 to $v + e = 4n^2 + 2n - 1$ occurs exactly once in C. (We know that 0 appears in the $(0, 0)$ position, as required.)

The row sums of C are:

$$\sum_{j=0}^{2n} c_{ij} = c_{i0} + \sum_{j=1}^{2n} a_{ij} + 2n \sum_{j=1}^{2n} b_{ij} + 2n^2$$

$$= c_{i0} + 2n^2 - i + 2n \sum_{j=1}^{2n} b_{ij} + 2n^2$$

$$= 4n^3 + 2n^2,$$

after inserting the appropriate values of c_{i0} and $\sum_{j=1}^{2n} b_{ij}$, depending on whether or not $i \leq n - 1$. Similarly, the column sums are:

$$\sum_{i=0}^{2n-1} c_{ij} = c_{0j} + \sum_{i=1}^{2n-1} a_{ij} + 2n \sum_{i=1}^{2n-1} b_{ij} + (2n - 1)n$$

$$= 4n^2 + 2n - j + 2n^2 - 3n + j$$

$$ + 2n(2n^2 - 3n + 1) + n(2n - 1)$$

$$= 4n^3 + 2n^2.$$

Thus all row and column sums (except the first) equal $4n^3 + 2n^2$, as required.

Finally, we prove that each integer from 0 to $4n^2 - 2n - 1$ appears exactly once in C. The numbers $0, 1, 2, \ldots, n - 1$ and $4n^2 - n, 4n^2 - n + 1, \ldots, 4n^2 + 2n - 1$ appear in the first row and column. The entries in A lie in the (closed) interval $[0, 2n - 1]$ and those of B lie in the interval $[0, 2n - 2]$. Thus the entries in C outside the first row and column lie in the interval $[n, 3n - 1 + 2n(2n - 2)] = [n, 4n^2 - n - 1]$. There are $4n^2 - 2n$ such integers so it remains to show that the entries are distinct. To prove this we need to show that the pairs (a_{ij}, b_{ij}) are distinct. The first and last columns of A contain only the integers 0 and $2n - 1$, the first and last columns of B contain the integers $0, 1, \ldots, 2n - 2$, and it is easy

to see that there are no repeated pairs. For the rest of the matrices A and B, note that the entries b_{ij} for a fixed value of $i + j$ mod $(2n - 1)$ are constant, while in matrix A the equivalent entries take all the values $1, 2, \ldots, 2n - 2$. Thus all pairs are distinct and so each integer from 1 to $4n^2 - 2n - 1$ appears exactly once in C.

\square

Labeling $K_{m,m+1}$, m even

In this case we write $m = 2n$. Then $v = 4n + 1$, $e = 4n^2 + 2n$, and a total labeling requires $4n^2 + 6n + 1$ labels.

Theorem 3.11 *There exists a vertex-magic total labeling of $K_{2n,2n+1}$ with magic constant $(n + 1)(2n + 1)^2$.*

Proof. We construct a representation matrix $C = (c_{ij})$ for a vertex-magic total labeling of $K_{2n,2n+1}$ as follows.

(i) Row 0 of C is $0, (2n+1)^2, (2n+1)^2+1, \ldots, (2n+1)^2+2n$, that is $c_{00} = 0$ and $c_{0j} = (2n + 1)^2 + j - 1$ for $1 \le j \le 2n + 1$.

(ii) $c_{i0} = (2n + 2)i, 1 \le i \le 2n$.

(iii) If $1 \le i < n$ and $1 \le j \le n+1$, or if $n+2 \le i \le 2n$ and $n+2 \le j \le 2n+1$, then
$$c_{ij} = 2n(2n + 2) - [j + (i - 1)(2n + 2)].$$

(iv) If $1 \le i < n$ and $n+2 \le j \le 2n+1$, or if $n+1 < i \le n$ and $1 \le j \le n+1$, then
$$c_{ij} = j + (i - 1)(2n + 2).$$

(v) If $1 \le j \le n + 1$, then
$$c_{nj} = n(2n + 2) + 2n - 2j + 3, \ c_{n+1,j} = (n - 1)(2n + 2) + n + j.$$

If $n + 2 \le j \le 2n + 1$, then
$$c_{nj} = (n - 1)(2n + 2) + 4n - 2j + 4 \ c_{n+1,j} = n(2n + 2) + j - n - 1.$$

Part (v) can also be expressed as follows: (except for column 0) rows 0, n and $n + 1$ of C are derived from rows of

$$X = \begin{bmatrix} 1 & 2 & 3 \ldots & n + 1 & n + 2 & n + 3 \ldots & 2n + 1 \\ 2n + 1 & 2n - 1 & 2n - 3 \ldots & 1 & 2n & 2n - 2 \ldots & 2 \\ n + 1 & n + 2 & n + 3 \ldots 2n + 1 & 1 & 2 \ldots & n \end{bmatrix}$$

by adding

$$\begin{bmatrix} (2n+1)^2 - 1 \\ n(2n+2) \\ (n-1)(2n+2) \end{bmatrix}$$

to each of the first $n+1$ columns and

$$\begin{bmatrix} (2n+1)^2 - 1 \\ (n-1)(2n+2) \\ n(2n+2) \end{bmatrix}$$

to the remainder. Each row of X is a permutation of $\{1, 2, \ldots, 2n+1\}$, so rows n and $n+1$ between them contain each of $(n-1)(2n+2)+1, (n-1)(2n+2)+2, \ldots, (n+1)(2n+2)$ exactly once. When $1 \le i < n$, rows i and $2n+1-i$ contain between them all the integers

$$t + (i-1)(2n+2) : 1 \le t \le 2n+2,$$
$$t + (2n-i)(2n+2) : 1 \le t \le 2n+2$$

precisely once each $(2n + 2 + (i-1)(2n+2) = i(2n+2) = c_{i0}, 2n+2+(2n-i)(2n+2) = c_{2n+1-i,0}$, and the others are given by (iii) and (iv)). Row 0 provides $0, (2n+1)^2, (2n+1)^2+1, \ldots, (2n+1)^2+2n = 4n^2+6n+1$. So C contains each of $0, 1, \ldots, 4n^2 + 6n + 1$ exactly once.

From (iii) and (iv), it also follows that

$$c_{ij} + c_{2n+1-i,j} = j + (i-1)(2n+2) + 2n(2n+2)$$
$$- [j + (i-1)(2n+2)]$$
$$= 2n(2n+2)$$

for $1 \le i < n$. Each column of X has sum $3n+3$, so $c_{nj} + c_{n+1,j} + c_{0,j} = (n+1)(8n+1)$ for $1 \le j \le 2n+1$. Therefore the sum of column j is

$$\sum_{i=0}^{2n} c_{ij} = (n-1)2n(2n+2) + c_{nj} + c_{n+1,j} + c_{0,j}$$
$$= (n-1)2n(2n+2) + (n+1)(8n+1)$$
$$= (n+1)(2n+1)^2.$$

If $i \le n$,

$$\sum_{j=0}^{2n+1} c_{ij} = \sum_{j=1}^{2n+1} j + c_{i0} + n(i-1)(2n+2) + (n+1)(2n-i)(2n+2)$$
$$= \binom{2n+2}{2} + (2n+2)i + (2n+2)[n(i-1) + (n+1)(2n-i)]$$
$$= (n+1)(2n+1) + 2(n+1)n(2n+1)$$
$$= (n+1)(2n+1)^2,$$

and a similar calculation gives the same sum if $i \ge n+1$. □

Remark. The array X is essentially the Kotzig array defined in (1.2). It is remarkable that Kotzig produced the array in solving an *edge*-magic total labeling problem, which did not involve complete bipartite graphs. This surprising connection was not noticed until after Theorem 3.11 had been proven, and it was then noticed that Kotzig's array had the required properties.

3.5.2 The Spectrum

In those cases where magic labelings are known to exist, it is interesting to know the set of values h such that there is a magic labeling with magic constant h. This is the *spectrum* of the labeling problem.

Suppose G has a vertex-magic total labeling λ. As usual, s_e is the sum of edge-labels: $s_e = \sum_{x \in E(G)} \lambda(x)$. Then counting the sum of labels at all vertices, we have from (3.3) and (3.4)

$$s_e + \binom{v + e + 1}{2} = vh, \tag{3.16}$$

$$\binom{e + 1}{2} \le s_e \le \binom{e + 1}{2} + ve. \tag{3.17}$$

These equations can be used to put bounds on the spectrum of vertex-magic total labelings. For $K_{m,m}$, (3.16) and (3.17) give

$$\frac{1}{2}[(m + 1)^3 - m^2] \le h \le \frac{1}{2}[(m + 1)^3 + m^2]. \tag{3.18}$$

As $K_{m,m}$ is regular, the duality Theorem, Theorem 3.2, applies, so there will be a vertex-magic total labeling with magic constant $\frac{1}{2}[(m + 1)^3 + x]$ if and only if there is one with $h = \frac{1}{2}[(m + 1)^3 - x]$.

For $K_{2,2}$, (3.18) yields $12 \le h \le 15$, and a complete search shows that every value can be realized. In fact, there are exactly six vertex-magic total labelings of $K_{2,2}$ (up to isomorphism). The representation matrices are:

$$
h = 12: \quad
\begin{bmatrix}
0 & 5 & 7 \\
\hline
8 & 1 & 3 \\
4 & 6 & 2
\end{bmatrix}
\qquad
h = 13: \quad
\begin{bmatrix}
0 & 7 & 3 \\
\hline
8 & 1 & 4 \\
2 & 5 & 6
\end{bmatrix}
\qquad
h = 13: \quad
\begin{bmatrix}
0 & 4 & 6 \\
\hline
7 & 1 & 5 \\
3 & 8 & 2
\end{bmatrix}
$$

$$
h = 14: \quad
\begin{bmatrix}
0 & 6 & 2 \\
\hline
7 & 3 & 4 \\
1 & 5 & 8
\end{bmatrix}
\qquad
h = 14: \quad
\begin{bmatrix}
0 & 5 & 3 \\
\hline
6 & 1 & 7 \\
2 & 8 & 4
\end{bmatrix}
\qquad
h = 15: \quad
\begin{bmatrix}
0 & 4 & 2 \\
\hline
5 & 3 & 7 \\
1 & 8 & 6
\end{bmatrix}.
$$

For $K_{3,3}$, the bounds are $28 \leq h \leq 36$, and all these values can be realized. There are 35 isomorphism classes with $h = 28$ and with $h = 36$, 70 with $h = 29$ and with $h = 35$, 477 with $h = 30$ and with $h = 34$, 250 with $h = 31$ and with $h = 33$, and 882 with $h = 32$.

In view of the ease with which examples are found for small m, we conjecture that every value of h allowed by (3.18) can be realized, but this is far from established.

Research Problem 3.2 *Prove or disprove that every value of h allowed by (3.18) can be realized as the magic constant of a vertex-magic total labeling of $K_{m,m}$.*

In the case of $K_{m,m+1}$ we can improve on (3.17) by an argument similar to that used in the proof of Theorem 3.8. Write s_1 and s_2 for the sums of the labels on the m-set and $(m+1)$-set of vertices, respectively, and s_e again for the sum of the edge-labels. Then every edge is adjacent to exactly one of the vertices in each set. Adding all labels on or adjacent to all vertices in the m-set, we get

$$hm = s_1 + s_e$$
$$\geq 1 + 2 + \cdots + (m + m(m+1)),$$

so

$$h \geq \tfrac{1}{2}(m+1)^2(m+2) \qquad (3.19)$$

while the larger set of vertices yields

$$h(m+1) = s_2 + s_e$$
$$\leq (m+1) + (m+2) + \cdots + ((2m+1) + m(m+1)),$$

whence

$$h \leq \tfrac{1}{2}(m+1)(m^2 + 4m + 2). \qquad (3.20)$$

For $K_{1,2}$, (3.19) and (3.20) yield $6 \leq h \leq 7$, and both values can be realized. In fact there are exactly two labelings up to isomorphism, with representation matrices

$$h = 6: \qquad h = 7:$$

$$\begin{bmatrix} 0 & 4 & 5 \\ 3 & 2 & 1 \end{bmatrix} \qquad \begin{bmatrix} 0 & 5 & 3 \\ 1 & 2 & 4 \end{bmatrix}.$$

However, for $K_{2,3}$, the bounds are $18 \leq h \leq 21$, but only 18, 19, and 20 can be realized. There are four labelings up to isomorphism:

$h = 18:$ $h = 19:$ $h = 19:$ $h = 20:$

$$
\begin{bmatrix} 0 & 11 & 10 & 9 \\ 7 & 4 & 6 & 1 \\ 5 & 3 & 2 & 8 \end{bmatrix}
\quad
\begin{bmatrix} 0 & 11 & 9 & 8 \\ 7 & 5 & 6 & 1 \\ 2 & 3 & 4 & 10 \end{bmatrix}
\quad
\begin{bmatrix} 0 & 11 & 9 & 8 \\ 5 & 6 & 7 & 1 \\ 4 & 2 & 3 & 10 \end{bmatrix}
\quad
\begin{bmatrix} 0 & 11 & 9 & 6 \\ 1 & 7 & 8 & 4 \\ 5 & 2 & 3 & 10 \end{bmatrix}.
$$

For $K_{3,4}$, labelings are easily found for each h satisfying the bounds $40 \leq h \leq 46$.

An obvious open question is: for which h satisfying (3.19) and (3.20) do vertex-magic total labelings exist? We lean toward the view that the case $m = 2, h = 21$ is a "small numbers" anomaly, and that all other possible magic constants can be realized.

Research Problem 3.3 *Prove or disprove that every value of h allowed by* (3.19) *and* (3.20) *can be realized as the magic constant of a vertex-magic total labeling of $K_{m,m+1}$.*

3.5.3 Joins

Complete bipartite graphs arise in the definition of joins of graphs. Suppose G and H are disjoint graphs. The *join* of G and H, denoted $G \vee H$, is the union of G, H and the complete bipartite graph with vertex-sets $V(G)$ and $V(H)$.

Suppose G and H are graphs that each have v vertices, and suppose the disjoint union $G \cup H$ has a vertex-magic total labeling λ with magic constant h. Suppose there exists a magic square A of size $v \times v$. The magic constant will be $\frac{1}{2}v(v^2 + 1)$. Define a labeling μ of $G \vee H$ as follows: for the vertices and edges of G and H, $\mu = \lambda$; if x is a vertex of G and y is a vertex of H, then $\mu(xy) = 2v + |E(G)| + |E(H)| + a_{xy}$. Then μ is easily seen to be a vertex-magic total labeling of $G \vee H$ with magic constant $h + v(2v + |E(G)| + |E(H)|) + \frac{1}{2}v(v^2 + 1)$. Since magic squares of all orders exist, we have:

Theorem 3.12 *If G and H are graphs of the same order such that the disjoint union $G \cup H$ has a vertex-magic total labeling, then $G \vee H$ has a vertex-magic total labeling.*

3.5.4 Unions of Stars

We have seen that $K_{1,2}$ is the only vertex-magic star. It is reasonable to ask whether disjoint unions of stars are vertex-magic.

Suppose G is the union $K_{1,n_1} \cup K_{1,n_2} \cup \ldots \cup K_{1,n_t}$ of t stars. Write N for $n_1 + n_2 + \cdots + n_t$. Suppose G has a vertex-magic total labeling with magic

constant h. The sum of the weights of the centers of the stars will be th; on the other hand, it will equal at least the sum of the smallest $N + t$ positive integers (the N spokes and the t centers). So

$$th \geq \sigma_0^{N+t} = \tfrac{1}{2}(N + t)(N + t + 1). \tag{3.21}$$

On the other hand, the sum of the N weights of the leaves equals the sum of the labels on all the edges and all the vertices except the centers, so

$$Nh \leq \sigma_t^{2N+t} = \tfrac{1}{2}(2N + t)(2N + t + 1) - \tfrac{1}{2}t(t + 1). \tag{3.22}$$

Combining (3.21) and (3.22),

$$
\begin{aligned}
N(N + t)(N + t + 1) &\leq t(2N + t)(2N + t + 1) - t^2(t + 1), \\
(N + t)(N + t + 1) &\leq N^2(4t) + N(4t^2 + 2t),
\end{aligned}
$$

so $N^2 + N(1 - 2t) - (3t^2 + t) \leq 0$. It follows that

$$N \leq \frac{2t - 1 + \sqrt{16t^2 + 1}}{2} \leq 3t. \tag{3.23}$$

Theorem 3.13 *If a disjoint union of stars is vertex-magic, then the average size of the component stars is less than* 3.

Exercise 3.7 *Find vertex-magic total labelings of* $2K_{1,2}$ *and* $K_{1,2} \cup K_{1,3}$.

It is clear that Theorem 3.13 is the best-possible conclusion from (3.23), because $\frac{2t-1+\sqrt{16t^2+1}}{2} > 3t - 1$. But not every union of stars with average size smaller than 3 is vertex-magic.

Suppose G is a vertex-magic union of t stars that has $3t - 1$ edges. Then (3.21) and (3.22) yield

$$8t - 2 \leq h \leq 8t - 1.$$

Consider any vertex x of degree 1. The label on the edge adjacent to x is no greater than $7t - 2$, so $\lambda(x) \geq t$. Therefore $1, 2, \ldots, t - 1$ must all be labels of centers of stars; and if $h = 8t - 1$ then t is also a center label.

Suppose one of the stars has s edges. Write c for the center of this star. If $h = 8t - 2$, then c has weight at least $1 + t + (t + 1) + \cdots + (t + s - 1) = 1 + st + \tfrac{1}{2}s(s - 1)$, so

$$8t - 2 \geq 1 + st + \tfrac{1}{2}s(s - 1). \tag{3.24}$$

Clearly $s < 8$, no matter what value t takes. Even for smaller s, not all t are possible. The inequality (3.24) can be written as

$$s^2 + (2t - 1)s + 6 - 16t \leq 0,$$

so

$$s \leq \frac{1 - 2t + \sqrt{4t^2 + 60t - 23}}{2}.$$

In the case $h = 8t - 1$, c has weight at least $1 + (t + 1) + \cdots + (t + s) = 1 + st + \frac{1}{2}s(s + 1)$. Again $s < 8$. For smaller s we obtain the slightly weaker condition

$$s \leq \frac{-1 - 2t + \sqrt{4t^2 + 68t - 15}}{2}.$$

From this we can deduce the following:

Theorem 3.14 *Suppose G is the vertex-magic union of t disjoint stars which between them have $3t - 1$ edges; then no star can contain 8 edges. If the largest star has s edges, then:*

$$\begin{aligned}
\text{if } s = 7 \quad &\text{then} \quad t \geq 30 \\
\text{if } s = 6 \quad &\text{then} \quad t \geq 12 \\
\text{if } s = 5 \quad &\text{then} \quad t \geq 6 \\
\text{if } s = 4 \quad &\text{then} \quad t \geq 3 \\
\text{if } s = 3 \quad &\text{then} \quad t \geq 2.
\end{aligned}$$

The two extreme cases are worth considering. If t is any positive integer, then $(t - 1)K_{1,3} \cup K_{1,2}$ is always a possibility according to the Theorem, and it is in fact vertex-magic. One labeling is shown in Fig. 3.5

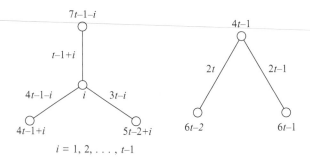

$$i = 1, 2, \ldots, t{-}1$$

Fig. 3.5. Vertex-magic total labeling of $(t - 1)K_{1,3} \cup K_{1,2}$

However, $(t - 1)K_{1,2} \cup K_{1,t+1}$ can never have a vertex-magic total labeling when $t > 3$. The cases $t = 2, 3$ are vertex-magic—when $t = 2$, the construction of Fig. 3.5 provides an example. The case $t = 3$ is left as an exercise.

Exercise 3.8 *Find a vertex-magic total labeling of $2K_{1,2} \cup K_{1,4}$.*

Research Problem 3.4 in our first edition asked, "Are the bounds in Theorem 3.14 the best possible?" In fact, Mihalisin [75] showed that those bounds are infact

sharp for the remaining cases. The examples are given below. Each set of internal parentheses denotes a star, beginning with its center label followed by its edge labels. Leaf labels are implied.

$\{(s = 5; h = 47)\{2 : 7, 8, 9, 10, 11\}, \{4 : 12, 14, 17\}, \{3 : 13, 15, 16\},$
$\quad \{1 : 20, 26\}, \{5 : 18, 24\}, \{6 : 19, 22\}\}$

$\{(s = 6; h = 95)\{1 : 13, 14, 15, 16, 17, 19\}, \{2 : 18, 32, 43\}, \{3 : 23, 28, 41\},$
$\quad \{4 : 21, 30, 40\}, \{5 : 22, 29, 39\}, \{7 : 24, 27, 37\}, \{8 : 25, 26, 36\},$
$\quad \{10 : 20, 31, 34\}, \{6 : 44, 45\}, \{9 : 33, 53\}, \{11 : 38, 46\}, \{12 : 35, 48\}\}$

$\{(s = 7; h = 239)\{1 : 31, 32, 33, 34, 35, 26, 37\}, \{2 : 38, 36, 113\},$
$\quad \{4 : 40, 83, 112\}, \{5 : 41, 82, 111\}, \ldots, \{25 : 61, 62, 91\}, \{26 : 39, 84, 90\},$
$\quad \{3 : 117, 119\}, \{28 : 87, 124\}, \{29 : 89, 121\}, \{30 : 86, 123\}, \{32 : 90, 117\}\}$

3.6 Vertex-Magic Total Labelings of Complete Tripartite Graphs

Complete tripartite graphs are more complicated than complete bipartite graphs. It is not known which of these graphs is vertex-magic, and Cattell [21] conjectures that no universal construction exists for labeling those $K_{k,m,n}$ that are vertex-magic. However, two families are known; the constructions come from [21].

Theorem 3.15 *There is a vertex-magic labeling of* $K_{1,n,n}$ *whenever n is odd; the magic constant is* $\frac{1}{2}(n^3 + 6n^2 + 9n + 2)$.

Proof. We denote the vertices of $K_{1,n,n}$ by $\{x_1, x_2, \ldots, x_n, y_1, y_2, \ldots, y_n, z\}$. The edges are $\{x_i y_j, x_i z, y_j z : i \leq i, j \leq n\}$. The total number of vertices and edges is $n^2 + 4n + 1$.

To construct the labeling, select a magic square M of side n, and define an array A by $a_{ij} = m_{ij} + 2n$. Then define

$$\begin{aligned}
\lambda(z) &= n^2 + 4n + 1, \\
\lambda(x_i) &= n^2 + 4n + 1 - i, \\
\lambda(y_j) &= n^2 + 3n + 1 - j \quad \text{if } j \text{ is odd}, \\
\lambda(y_j) &= 2n + 1 - j \quad\quad\;\; \text{if } j \text{ is even}, \\
\lambda(x_i y_j) &= a_{ij}, \\
\lambda(x_i z) &= i, \\
\lambda(y_j z) &= n + j \quad\quad\quad\; \text{if } j \text{ is odd}, \\
\lambda(y_j z) &= n^2 + 2n + j \quad\; \text{if } j \text{ is even}.
\end{aligned}$$

It is easy to check that this is a vertex-magic labeling with the required magic constant. \square

Exercise 3.9 *Verify that the labeling given in Theorem 3.15 is in fact vertex-magic.*

Theorem 3.16 *There is a vertex-magic labeling of $K_{2,n,n}$ whenever $n \equiv 3$ (mod 4).*

Proof. Assume $n \geq 15$. We denote the vertices by $\{x_1, x_2, \ldots, x_n, y_1, y_2, \ldots, y_n, z_1, z_2\}$, and the edges in the obvious way.

Define the vertex labels by

$$
\begin{aligned}
\lambda(x_i) &= 3n + 3 - 2i & &\text{for } i = 1, \ldots, \tfrac{1}{2}(n - 11), \\
\lambda(x_i) &= 2n + 3 - 2i & &\text{for } i = \tfrac{1}{2}(n - 9), \ldots, \tfrac{1}{2}(n - 3), \\
\lambda(x_{(n-1)/2}) &= n + 3, \\
\lambda(x_{(n+1)/2}) &= n + 2, \\
\lambda(x_i) &= n^2 + 2n + 4i, & &\text{for } i = \tfrac{1}{2}(n + 3), \ldots, n - 1, \\
\lambda(x_n) &= n^2 + 6n - 2, \\
\lambda(y_j) &= 4n + 3 - 2j & &\text{for } j = 1, \ldots, \tfrac{1}{2}(n - 5), \\
\lambda(y_{(n-3)/2}) &= 5, \\
\lambda(y_{(n-1)/2}) &= 2, \\
\lambda(y_j) &= n^2 + 2n + 1 + 4j, & &\text{for } j = \tfrac{1}{2}(n + 1), \ldots, n - 1, \\
\lambda(y_n) &= n^2 + 6n - 1, \\
\lambda(z_1) &= 1, \\
\lambda(z_2) &= 3n + 2.
\end{aligned}
$$

Now select a magic square M of side n, and define an array A by $a_{ij} = m_{ij} + 4n + 2$. Set $\lambda(x_i y_j) = a_{ij}$. The other edge labels involving z_1 are

$$
\begin{aligned}
\lambda(x_i z_1) &= n^2 + 4n + 4i & &\text{for } i = 1, \ldots, \tfrac{1}{2}(n - 3), \\
\lambda(x_{(n-1)/2} z_1) &= n^2 + 6n, \\
\lambda(x_{(n+1)/2} z_1) &= n^2 + 6n + 2, \\
\lambda(x_i z_1) &= 3n + 3 - 2i & &\text{for } i = \tfrac{1}{2}(n + 3), \ldots, n - 2, \\
\lambda(x_{n-1} z_1) &= 2n + 5, \\
\lambda(x_n z_1) &= 2n + 4, \\
\lambda(y_j z_1) &= n + 2 - 2j & &\text{for } j = 1, \ldots, \tfrac{1}{2}(n - 5), \\
\lambda(y_{(n-3)/2} z_1) &= 3n + 6, \\
\lambda(y_{(n-1)/2} z_1) &= 3n + 3, \\
\lambda(y_j z_1) &= 5n + 3 - 2j & &\text{for } j = \tfrac{1}{2}(n + 1), \ldots, \tfrac{1}{4}(3n - 5), \\
\lambda(y_j z_1) &= 2n + 2 - 2j & &\text{for } j = \tfrac{1}{4}(3n - 1), \ldots, n - 1, \\
\lambda(y_n z_1) &= 3,
\end{aligned}
$$

while those involving z_2 are

$$
\begin{aligned}
\lambda(x_i z_2) &= 2n + 3 - 2i & \text{for } i = 1, \ldots, \tfrac{1}{2}(n - 11), \\
\lambda(x_i z_2) &= 3n + 3 - 2i & \text{for } i = \tfrac{1}{2}(n - 9), \ldots, \tfrac{1}{2}(n - 3), \\
\lambda(x_{(n-1)/2} z_2) &= 2n + 3, \\
\lambda(x_{(n+1)/2} z_2) &= 2n + 2, \\
\lambda(x_i z_2) &= 4n + 3 - 2i & \text{for } i = \tfrac{1}{2}(n + 3), \ldots, n - 2, \\
\lambda(x_{n-1} z_2) &= n + 5, \\
\lambda(x_n z_2) &= n + 4, \\
\lambda(y_j z_2) &= n^2 + 4n + 1 + 4j & \text{for } j = 1, \ldots, \tfrac{1}{2}(n - 5), \\
\lambda(y_{(n-3)/2} z_2) &= n^2 + 6n - 5, \\
\lambda(y_{(n-1)/2} z_2) &= n^2 + 6n + 1, \\
\lambda(y_j z_2) &= 2n + 2 - 2j & \text{for } j = \tfrac{1}{2}(n + 1), \ldots, \tfrac{1}{4}(3n - 5), \\
\lambda(y_j z_2) &= 5n + 3 - 2j & \text{for } j = \tfrac{1}{4}(3n - 1), \ldots, n - 1, \\
\lambda(y_n z_2) &= 3n + 4.
\end{aligned}
$$

A check shows that each vertex has weight $\tfrac{1}{2}(n^3 + 10n^2 + 23n + 12)$.

This construction does not work for $n = 3, 7, 11$, so these cases must be constructed by brute force computation, but all exist. □

Exercise 3.10 *Verify that the labelings given in Theorems* 3.15 *and* 3.16 *are in fact vertex-magic.*

It may be shown that if the graph $K_{k,m,n}$ is vertex-magic, then $k + m \geq n - 1$. This result is best-possible, as some examples are known where $k + m = n - 1$ and $K_{k,m,n}$ is vertex-magic (constructions are given in [21] for vertex-magic labelings of $K_{1,1,3}$ and $K_{1,2,4}$).

3.7 Graphs with Vertices of Degree One

For a graph to be vertex-magic, the presence of degree one vertices turns out to create a restriction on both the number of edges and the number of vertices of higher degree in the graph. As an illustration, we first examine a family of graphs in which the number of vertices of degree one is the same as the number of vertices of higher degree.

Let G be any graph of order v and size e. We define a *G-sun* to be a graph G^* of order $2v$ formed from G by adjoining v new vertices of degree 1, one to each vertex of G. We have $v^* = |V(G^*)| = 2v$ and $e^* = |E(G^*)| = e + v$. When G is the cycle C_v and a new vertex is adjoined to each vertex of G, the resulting G^* is a *sun* as defined in Sect. 2.4.2.

Let us call the edges of G the *inner edges* of the G-sun and the vertices of G the *inner vertices*; the other elements (leaves and their adjacent edges) are the *outer edges* and *outer vertices*. The following theorem shows that for a labeling to exist, the number of edges in G must be bounded above by a function of v that is essentially linear.

Theorem 3.17 *Let G be any graph of order v. If G has e edges, then a G-sun G^* is not vertex-magic total when*

$$e > \frac{-1 + \sqrt{1 + 8v^2}}{2}$$

Proof. The label set for G^* is $\{1, 2, \ldots, 3v + e\}$. We calculate the minimum possible sum of weights on the inner vertices; this is achieved by putting the e smallest labels on the inner edges and the next $2v$ smallest labels on the inner vertices and outer edges. Remembering that the inner edge labels will each be added twice, this gives us

$$\sum \text{wt}(x_i) \geq \sum_1^{2v+e} i + \sum_1^e i$$
$$= \tfrac{1}{2}\left[(2v + e)(2v + e + 1) + e(e + 1)\right]$$

Since there are v inner vertices, we must therefore have:

$$h \geq \frac{1}{v}\left[2v^2 + v(2e + 1) + e(e + 1)\right].$$

Calculating the maximum possible sum of weights on the outer vertices, we find (taking the sum of the $2v$ largest labels):

$$\sum \text{wt}(x_i) \leq \sum_1^{3v+e} i - \sum_1^{v+e} i$$
$$= \tfrac{1}{2}\left[(3v + e)(3v + e + 1) - (v + e)(v + e + 1)\right]$$
$$= 4v^2 + 2ve + v.$$

Since there are v outer vertices, we have

$$h \leq \tfrac{1}{v}(4v^2 + 2ve + v)$$
$$= 4v + 1 + 2e.$$

Consequently, a labeling cannot exist whenever

$$\frac{1}{v}\left[2v^2 + v(2e + 1) + e(e + 1)\right] > 4v + 1 + 2e.$$

This simplifies to

$$e^2 + e - 2v^2 > 0$$

and the theorem follows. □

A second illustration of the impact of degree 1 vertices is found by examining trees. This time the existence of a vertex-magic total labeling forces a lower bound on the number of internal vertices in the graph. We have the following :

Theorem 3.18 *Let T be a tree with n internal vertices. If T has more than $2n$ leaves, then T does not admit a vertex-magic total labeling.*

Proof. Suppose T has n internal vertices $\{x_1, x_2, \ldots, x_n\}$ and $2n + t$ leaves $\{y_1, y_2, \ldots, y_{2n+t}\}$, so that $v = 3n + t$ and $e = 3n + t - 1$. Then the label set is $\{1, 2, \ldots, 6n + 2t - 1\}$.

The maximum possible sum of weights on the leaves is the sum of the $4n + 2$ largest labels:

$$\sum \mathrm{wt}(y_i) \le \sum_{1}^{6n+2t-1} i - \sum_{1}^{2n-1} i$$
$$= (8n + 2t - 1)(2n + t)$$

and since there are $2n + t$ leaves, $h \le 8n + 2t - 1$. The minimum possible sum of weights on the internal vertices is the sum of all but the $2n + t$ largest labels:

$$\sum \mathrm{wt}(x_i) \ge \sum_{1}^{4n+t-1} i$$
$$= \tfrac{1}{2}(4n + t)(4n + t - 1).$$

Since there are n internal vertices,

$$h \ge \frac{(4n + t)(4n + t - 1)}{2n}$$

and these two inequalities imply

$$(4n + t)(4n + t - 1) \le 2n(8n + 2t - 1)$$

and

$$2n(2t - 1) + t(t - 1) \le 0. \tag{3.25}$$

Since n is at least 1, t cannot be positive. □

The labeling of the tree with two internal vertices and four leaves displayed in Fig. 3.6 shows that the statement of the theorem provides the best possible general result for trees.

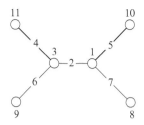

Fig. 3.6. Theorem 3.18 is best possible

Research Problem 3.4 *For each positive integer n, find a tree with n central vertices and 2n leaves, which is vertex-magic.*

3.8 The Complete Graphs

A systematic search for vertex-magic total labelings of K_4 has revealed interesting results. In this case $\lambda_i \in \{1, 2, \ldots, 10\}$ and (3.4) imply that $19 \leq h \leq 25$. The table shows the number of distinct labelings for each value of h.

h	$=$ 19	none
h	$=$ 20	2
h	$=$ 21	5
h	$=$ 22	none
h	$=$ 23	5
h	$=$ 24	2
h	$=$ 25	none

As described in Sect. 3.2, these come in dual pairs also: those labelings with $h = 24$ are the duals of those with $h = 20$ and those with $h = 23$ are duals of those with $h = 21$. The graph labeling of $K_4 - e$ given in Fig. 3.1 was obtained from one of the above labelings with $h = 20$ by deleting an edge labeled with 1, as described in Theorem 3.3.

We can apply equation (3.4) for arbitrary v to find the range of feasible values of h for K_v. Using $e = \frac{v(v-1)}{2}$, we find:

$$\frac{v(v^2 + 3)}{4} \leq h \leq \frac{v(v + 1)^2}{4}.$$

Theorem 3.19 [40, 58, 61] *There is a vertex-magic total labeling of K_v for all v.*

Proof. We shall use two $n \times n$ matrices, where n is odd. A denotes the matrix formed from the row $1, 2, \ldots, v$ by back-circulating—$a_{i,j} = a_{i-1,j+1}$, with

subscripts reduced modulo v when necessary—so that A is symmetric, and its diagonal entries are all different. If S is the sequence $s_0, s_1, s_2, \ldots, s_{\frac{1}{2}(n-1)}$, then $B(S) = (b_{i,j})$ is the matrix formed by circulating the row

$$s_0, s_1, s_2, \ldots, s_{\frac{1}{2}(n-1)}, s_{\frac{1}{2}(n-1)}, \ldots, s_1$$

—in other words, $b_{1,1} = s_0, b_{1,2} = s_1, \ldots, b_{1,n} = s_1$, and $b_{i,j} = b_{i-1,j-1}$, with subscripts taken modulo v when necessary.

Now define a labeling $\lambda_S(n)$ of K_n by $\lambda(x_i) = a_{i,i} + b_{i,i}$, $\lambda(x_i x_j) = a_{i,j} + b_{i,j}$. It is easy to see that under this labeling, every vertex x has the same weight:

$$\mathrm{wt}(x) = s_0 + 2(s_1 + s_2 + \cdots + s_{\frac{1}{2}(n-1)}) + 1 + 2 + \cdots + n.$$

In the case where v is odd we take $v = n$ and use the sequence

$$S = (0, n, 2n, \ldots, \tfrac{1}{2}n(n-1)).$$

Every label from 1 to $\frac{1}{2}n(n+1)$ will occur exactly once, so we have a vertex-magic total labeling of K_n. So there is a vertex-magic total labeling of K_n whenever n is odd.

If $v \equiv 2 \pmod 4$, we write $v = 2n$. We find a vertex-magic total labeling of the union of two copies of K_n. Then K_v is the join of these two copies of K_n, so Theorem 3.12 provides a vertex-magic total labeling of K_v.

To label $2K_n$ we distinguish two subcases. If $n = 4m + 1$, consider the two sequences

$$
\begin{aligned}
S_1 &= 2mn, 0, 2n, \ldots, (2m-2)n, (2m+3)n, \\
&\quad (2m+5)n, \ldots, (4m+1)n, \\
S_2 &= (2m+2)n, n, 3n, \ldots, (2m+1)n, (2m+4)n, \\
&\quad (2m+6)n, \ldots, 4mn.
\end{aligned}
$$

$\lambda_{S_1}(n)$ and $\lambda_{S_2}(n)$ can each be used to label K_n. Each has magic constant $(2m+1)(4m+1)^2$ and between them their sets of labels make up all the integers from 1 to $2\binom{n+1}{2}$. If these labelings are applied to two disjoint copies of K_n, they make up a vertex-magic total labeling of $2K_n$ as required.

In the same way, if $n = 4m + 3$, the sequences

$$
\begin{aligned}
S_1 &= 2m, 0, 2, \ldots, 2m-2, 2m+2, 2m+5, 2m+7, \ldots, 4m+3, \\
S_2 &= 2m+4, 1, 3, \ldots, 2m+3, 2m+6, 2m+8, \ldots, 4m+2
\end{aligned}
$$

can be used to label $2K_n$.

If $v \equiv 0 \pmod 4$, say $v = 4m$, we treat K_{4m} as $K_{4m-3} \cup K_3$ with edges joining the two vertex-sets. The copy of K_{4m-3} is labeled using $\lambda_S(4m - 3)$, where

$$S = 4m, 0, 8m - 3, 12m - 6, 16m - 9, \ldots,$$
$$(8m - 3) + (m - 3)(4m - 3), 8m + (m + 1)(4m - 3),$$
$$8m + (m + 2)(4m - 3), \ldots, 8m + (2m - 1)(4m - 3),$$

yielding constant vertex-weight $(2m + 1)(8m^2 - 6m - 3)$. The vertices of K_3 receive labels $4m - 2, 4m - 1, 4m$, and the edges receive $8m - 2 + (m - 2)(4m - 3), 8m - 1 + (m - 2)(4m - 3), 8m + (m - 2)(4m - 3)$, in such a way as to give each of the three vertices weight $8m^2 - 2m + 9$. Finally, a magic rectangle R of size $3 \times (4m - 3)$ is chosen, and the cross-edge joining vertex i of K_3 to vertex j of K_{4m-3} is labeled $8m + (m - 2)(4m - 3) + r_{ij}$. The magic rectangle has row and column sums $(4m - 3)(6m - 4)$ and $3(6m - 4)$, so the sum on each vertex of K_{4m-3} of the labels on the cross-edges is $3[(6m-4)+8m+(m-2)(4m-3)]$, and for the vertices of K_3 it is $(4m-3)[(6m-4)+8m+(m-2)(4m-3)]$. Therefore the combined labeling gives constant vertex-weight $16m^3 + 8m^2 - 3m + 3$. Every integer from 1 to $2m(4m+1)$ is used precisely once, so the result is a vertex-magic total labeling. □

Exercise 3.11 *Verify the computations of vertex-weight in the case $v \equiv 0 \pmod 4$, and verify that every positive integer up to $2n(4n + 1)$ appears precisely once among the labels.*

For small values of v, the following result has been established by brute force computation:

Theorem 3.20 *There is at least one vertex-magic total labeling for K_5 and for K_6 for every feasible value of h.*

The fact that vertex-magic total labelings seem to be plentiful for $n = 5$ and $n = 6$ suggests the following conjecture in our first edition: For each $n > 4$ there is a vertex-magic total labeling for K_n for every feasible value of h. This was proven for odd n by McQuillan and Smith [71]. Some other partial results for even n appear in [3].

3.9 Disconnected Graphs

Vertex-magic total labelings of disconnected graphs have not been widely studied. The following construction is a modification of the edge-magic total labeling construction in Sect. 2.8.4.

3.9.1 Multiples of Regular Graphs

By analogy with vertex colorings, a *C-edge-coloring* χ of a graph G is a map

$$\chi : E(G) \to C$$

to some set $C = \{c_1, c_2, \ldots\}$ of colors. A *proper edge-coloring* of G is an edge-coloring in which no two edges with a common endpoint belong to the same color class. In other words, if $x \sim y$ and $y \sim z$,

$$\chi(xy) \neq \chi(yz).$$

If $|C| = k$, the phrase *k-edge-coloring* is used.

Vizing [91] proved

Lemma 3.21 *Suppose the maximum degree among the vertices of G is Δ. Then G has a $(\Delta + 1)$-edge-coloring.*

(In fact, Vizing proved that the minimum number of colors needed is either Δ or $\Delta + 1$.)

Suppose G is a regular graph of degree Δ. Then G has maximum degree Δ, and can be properly edge-colored in $\Delta + 1$ colors. Choose such a coloring, χ say. At each vertex x, there will be exactly one color not represented on any edge adjacent to x. Define $\eta(x)$ to be that color. Define η to be the same as χ when applied to edges. Then η is a $(\Delta + 1)$-total coloring of G. We have proven:

Lemma 3.22 *Every regular graph of degree Δ has a total coloring in $\Delta + 1$ colors.*

Theorem 3.23 *Suppose G is a regular graph of degree Δ which has a vertex-magic total labeling.*

 (i) *If Δ is even, then nG is vertex-magic whenever n is an odd positive integer.*

 (ii) *If Δ is odd, then nG is vertex-magic for every positive integer n.*

Proof. Suppose G has e edges and v vertices. We wish to construct a vertex-magic total labeling of the ne-edge, nv-vertex graph $G_1 \cup G_2 \cup \ldots \cup G_n$, where each G_j is an isomorphic copy of G.

Select a vertex-magic total labeling λ of G, with magic constant h, and a Kotzig array $A = (a_{ij})$ of size $(\Delta + 1) \times n$, as guaranteed by Lemma 1.2. Let η be a total coloring of G in the $\Delta + 1$ colors $1, 2, \ldots, (\Delta + 1)$. We define a labeling Λ as follows. The element (vertex or edge) x_j of G_j corresponding to element x of G receives label $\lambda(x) + (e + v)a_{\eta(x),j}$.

Consider the elements $x_j, 1 \leq j \leq n$. These n elements receive the n labels $\eta(x), \eta(x) + e + v, \ldots, \eta(x) + (n-1)(e+v)$. These labels are precisely those integers in the range from 1 to $n(e+v)$ that are congruent to $\eta \pmod{(e+v)}$. As x ranges through the $e + v$ elements of G, each congruence class is represented exactly once. So Λ assumes each of the required values exactly once.

Now consider the labels on x_i and its incident edges. They are the labels on the corresponding elements of G, with $0, (e+v), \ldots, (n-1)(e+v)$ added to them in some order. So the sum is $h + (e+v)\binom{n}{2}$, which is constant, as required. □

Research Problem 3.5 *Can the method of Theorem 3.23 be applied to some case $G \cup H$ where G and H are nonisomorphic regular graphs?*

The case of a union of an even number of triangles is considered in [70], where it is shown that nC_3 is vertex-magic for every even integer n greater than 2. The construction is similar to that in the edge-magic case (see Sect. 2.8.2). It is also shown in [70] that for each even integer n greater than 4, nC_3 has vertex-magic total labelings with at least $2n - 2$ different magic constants.

3.10 Super Vertex-Magic Total Labelings

This section will explore super vertex-magic total labelings. These are special cases of vertex-magic total labelings in which either the edges are given the smallest possible labels (E-super) or the vertices are given the smallest possible labels (V-super).

3.10.1 E-super Vertex-Magic Total Labelings

The main results in this section are from [48] unless otherwise noted. Recall that a graph has an E-super vertex-magic total labeling λ if the labels $1, 2, \ldots, e$ are placed on the edges. In this case, the magic constant

$$h = e + \frac{v+1}{2} + \frac{e(e+1)}{v} \tag{3.26}$$

as

$$
\begin{aligned}
vh &= \sum_{u \in V} \lambda(u) + 2 \sum_{e \in E} \lambda(e) \\
&= (e+1) + (e+2) + \cdots + (e+v) + 2(e(e+1)/2) \\
&= ev + v(v+1)/2 + e(e+1).
\end{aligned}
$$

In addition, the following theorem is helpful in deciding when a particular graph has an E-super vertex-magic labeling.

Theorem 3.24 *Let G be a graph and γ is a bijection from E onto $\{1, 2, \ldots, e\}$. Then γ can be extended to an E-super vertex-magic labeling of G if and only if $\{w(u) = \sum_{v \in N(u)} \gamma(uv) | u \in V(G)\}$ consists of $|V(G)|$ consecutive integers.*

Proof. Assume that $\{w(u) | u \in V(G)\}$ consists of $|V(G)|$ consecutive integers. Let $t = \min\{w(u) | u \in V(G)\}$. Define $\lambda : V(G) \cup E(G) \to \{1, 2, \ldots, v+e\}$ as $\lambda(xy) = \gamma(xy)$ for $xy \in E(G)$ and $\lambda(x) = t + v + e - w(x)$. Then, $\lambda(E(G)) = \{1, 2, \ldots, e\}$ and $\lambda(V(G)) = \{e+v, e+v-1, \ldots, e+1\}$. Hence, λ is an E-super vertex-magic labeling with $h = t + v + e$.

Suppose γ can be extended to an E-super vertex-magic labeling λ of G with magic constant h. Now, let $t = \min\{w(u)\ u \in V(G)\}$. Since for every $u \in V(G)$, $\lambda(u) + w(u) = h$, we have $w(u) = h - \lambda(u)$. Thus, $\{w(u) | u \in V(G)\} = \{h - e - v, h - e - v + 1, \ldots, h - e - 1\} = \{t, t+1, \ldots, t+v-1\}$. $\qquad\square$

The following theorem is useful in finding classes of graphs that are not E-super vertex-magic.

Theorem 3.25 [9] *Let G be a graph with v vertices and e edges. If v is even and $e = v - 1$ or v, then G is not E-super vertex-magic.*

Proof. Suppose there exist an E-super vertex-magic labeling of G with magic constant h. Then, $h = e + \frac{v+1}{2} + \frac{e(e+1)}{v}$. Suppose $e = v - 1$ then

$$
\begin{aligned}
h &= v - 1 + \frac{v+1}{2} + \frac{(v-1)v}{v} \\
&= v - 1 + \frac{v+1}{2} + v - 1 \\
&= 2v - 2 + \frac{v+1}{2}, \quad \text{which is not an integer}
\end{aligned}
$$

Suppose $e = v$, then

$$
\begin{aligned}
h &= v + \frac{v+1}{2} + \frac{v(v+1)}{v} \\
&= v + \frac{v+1}{2} + v + 1 \\
&= 2v + 1 + \frac{v+1}{2}, \quad \text{which is not an integer}
\end{aligned}
$$

Hence G is not E-super vertex magic. $\qquad\square$

Corollary 3.25.1 *If v is even, then every tree T is not E-super vertex magic.*

Corollary 3.25.2 *A path P_n is E-super vertex-magic if and only if n is odd and $n \geq 3$.*

Exercise 3.12 *Find an E-super vertex-magic labeling of P_n for n odd and $n \geq 3$ that completes the proof of Corollary 3.25.2.*

Corollary 3.25.3 *A star graph $K_{1,n}$ is E-super vertex-magic if and only if $n = 2$.*

Proof. When $n = 2$, $K_{1,2}$ is P_3 which is E-super vertex-magic by Corollary 3.25.2.

Suppose $K_{1,n}$ is E-super vertex-magic. Then, if c is the central vertex, $w(c) = \frac{n(n+1)}{2}$. But, by Theorem 3.24 $w(c) = n + 1$ as the weights of the other vertices will be $\{1, 2, \ldots, n\}$. Thus, $n + 1 = \frac{n(n+1)}{2}$ but this only occurs if $n = 2$. □

Now consider the fan graphs, F_n. Suppose there exists an E-super vertex magic labeling for F_n with magic constant h. Then, $h = 2n - 1 + \frac{n+2}{2} + \frac{(2n-1)(2n)}{n+1} = \frac{13n^2+n}{2n+2}$ which is not an integer for all n except 2, 3, and 11.

Exercise 3.13 *Find an E-super vertex magic labeling for F_2. Then, show why there cannot be an E-super vertex magic labeling for F_3 and F_{11}.*

This gives,

Theorem 3.26 *The fan graph F_n is E-super vertex-magic if and only if $n = 2$.*

Corollary 3.26.1 *A cycle C_n is E-super vertex-magic if and only if n is odd.*

Proof. All that remains is to find the E-super vertex-magic labeling of C_n. Choose a vertex as v_1. Begin at vertex v_1 and label the first edge one. Continue around the cycle in a clockwise fashion labeling every other edge with the next consecutive integer. Once the last label has been placed on edge $v_{n-1}v_n$, place the next consecutive integer $(n + 1)$ on vertex v_n and move in a counter-clockwise fashion labeling each vertex with the next consecutive integer. □

Unlike the previous theorems about specific classes of graphs, the next two theorems give more general results on classes of graphs that could have an E-super vertex magic labeling and can be found in [9].

Theorem 3.27 *Let G be a graph of odd order. If G can be decomposed into two Hamiltonian cycles, then G is E-super vertex magic.*

Proof. Assume that G has v vertices, the length of the Hamiltonian cycle is odd say $v = 2m + 1$. Thus, the number of edges in G is $e = 4m + 2$. We choose a vertex v_1 and label the edges of the first Hamilton cycle starting at v_1 and moving in the clockwise direction by

$$\frac{v+1}{2}, 1, \frac{v+3}{2}, 2, \frac{v+5}{2}, \ldots, \frac{v-1}{2}, v.$$

Then we label the edges of the second Hamilton cycle starting at v_2 with counter clockwise direction by

$$\frac{3v+1}{2}, v+1, \frac{3v+2}{2}, v+2, \ldots, \frac{3v-1}{2}, 2v.$$

Then the sum of the edge labels at all the vertices are of v consecutive integers. Then, by Theorem 3.24, the result follows. \square

A stronger result based on decomposing graphs is given here without proof, but it is easy to construct the map that would produce the required E-super vertex-magic labeling.

Theorem 3.28 *If a graph G can be decomposed into two E-super vertex-magic spanning subgraphs G_1 and G_2 where G_2 is regular, then G is E-super vertex-magic.*

Other graphs that have been shown to be E-super vertex-magic include the complete graphs for n odd and (n, t)-kites if and only if $n + t$ is odd. The former can be proven in two ways. Since McQuillan and Smith showed that all feasible values in the magic spectrum can be obtained for odd complete graphs, we know there is an E-super vertex-magic labeling [71]. Also, Marimuthu and Balakrishnan give another method for finding E-super vertex-magic labelings for odd complete graphs [9]. The latter was shown in [49] and involves multiple cases based on the parity of n and t. The only disconnected graphs that have been shown to be E-super vertex magic are mC_n if and only if both m and n are odd [48].

3.10.2 V-super Vertex-Magic Labelings

The main results in this subsection are from [62] unless otherwise noted. For a V-super vertex-magic labeling of a graph G, the vertices receive the smallest labels. In other words if λ is a V-super vertex-magic labeling, then $\lambda(V) = \{1, 2, \ldots, v\}$. Note this implies the edges receive the largest labels possible and hence the magic constant is

$$h = \frac{(v+e)(v+e+1)}{v} - \frac{v+1}{2}. \tag{3.27}$$

Knowing the magic constant has to be an integer, we get the following corollary.

Corollary 3.28.1 *If G has a V-super vertex-magic labeling, then if v is odd, $v|e(e+1)$ and if v is even, $v|2e(e+1)$.*

The following theorem eliminates many families of graphs including one very important family given in the subsequent corollary.

Theorem 3.29 *If G has a vertex of degree 1, then G is not V-super vertex-magic.*

Proof. Let G be a graph of order $v \geq 3$ and let x_0 be the vertex of degree 1 in G. Then x_0 has a unique neighbor x of some degree $\delta > 1$. Let $e_0, e_1, \ldots, e_{\delta-1}$ be the edges incident with x where $e_0 = xx_0$. Suppose that G has a V-super VMTL λ. Then,

$$k = \lambda(x_0) + \lambda(e_0) = \lambda(x) + \lambda(e_0) + \cdots + \lambda(e_{\delta-1}).$$

Rearranging, we get

$$\lambda(x_0) - \lambda(x) = \lambda(e_1) + \cdots + \lambda(e_{\delta-1}).$$

Since λ is V-super, all terms on the right side must be larger than v, whereas the terms on the left side are each smaller than or equal to v. This is clearly a contradiction, so that G does not have a V-super vertex-magic labeling. \square

Corollary 3.29.1 *No tree admits a V-super vertex-magic labeling.*

Theorem 3.30 *The cycle C_n has a V-super vertex-magic labeling if and only if n is odd.*

Proof. Let $V(C_n) = \{v_1, v_2, \ldots, v_n\}$. Let $\lambda : V \cup E \to \{1, 2, \ldots, v+e\}$ be defined as $\lambda(v_i) = i$, $i = 1, 2, \ldots, n$. Label the edge $v_i v_{i+1}$ with

$$\lambda(v_i v_{i+1}) = \begin{cases} 2n - \frac{i-1}{2} & \text{if } i \text{ is odd} \\ 2n - \frac{n-1}{2} - \frac{i}{2} & \text{if } i \text{ is even} \end{cases}$$

It is easily checked that λ is a V-super vertex-magic labeling. It is easy to show that if n is even that the magic sum cannot be an integer. \square

Exercise 3.14 *Show that no ladder $(P_2 \times P_n)$, fan, or friendship graph can have a V-super vertex-magic labeling.*

It was shown in Sect. 3.5 that the only complete bipartite graphs with a vertex-magic total labeling are $K_{m,m}$ and $K_{m,m+1}$ for $m > 1$. Thus, these are the only complete bipartite graphs that could admit a V-super vertex-magic labeling. However, we can show that this is impossible in both cases.

Suppose that $K_{m,m}$ has a V-super vertex-magic labeling. Then, according to equation (3.27), the value of h is

$$h = \frac{(2m + m^2)(2m + m^2 + 1)}{2m} - \frac{2m + 1}{2}$$

and then

$$h = \frac{(m + 2)(2m + m^2 + 1) - 2m^2 - 1}{2}.$$

This value is not an integer for either odd or even m, and so it follows that $K_{m,m}$ cannot have a V-super vertex-magic labeling. For $K_{m,m+1}$ on the other hand, the number of edges is $e = m^2 + m$ while the number of vertices is $2m + 1$. The number of edges is not a multiple of v so, by Corollary 3.28.1, no V-super vertex-magic labeling exists. Thus, we have the following theorem.

Theorem 3.31 *No complete bipartite graph is V-super vertex-magic.*

All odd complete graphs are V-super vertex-magic. The results of McQuillan and Smith give vertex-magic labelings for all possible values in the magic spectrum, so this will certainly include the case when the magic sum is as large a possible [71]. If a complete graph has $n \equiv 2 \pmod 4$ vertices, then by Corollary 3.28.1 K_n will not have a V-super vertex magic labeling. In 2004, MacDougall et al. [62] conjectured that if $n \equiv 0 \pmod 4$ and $n > 4$, then K_n has a V-super vertex-magic total labeling. This conjecture was verified in 2008 by Gómez [34] who constructed a new labeling for these graphs which is V-super vertex-magic.

In [35], two new methods to obtain V-super vertex-magic total labelings of graphs are given. The first method uses two other types of labelings to create a V-super vertex-magic labeling of certain classes of kG where G is a regular graph. Gómez uses this result to show that kK_n has a V-super vertex-magic labeling if n and k are odd or $n = 4s$ where $s = 2, 3, 4, \ldots$. The second method takes a particular graph that has a V-super vertex-magic labeling and adds a certain collection of edges to produce a new graph that also has a V-super vertex-magic labeling.

Exercise 3.15 *Show that a regular graph has a V-super vertex-magic total labeling if and only if it has an E-super vertex-magic total labeling.*

3.11 Vertex-Magic Injections

A vertex-magic injection is like a vertex-magic total labeling, except that the labels can be any positive integers. As in the edge-magic case, we define an $[m]$-vertex-magic injection of G to be a vertex-magic injection of G in which

the largest label is m, and again we call m the *size* of the injection. If G has a vertex-magic injection, the *vertex deficiency* $def_v(G)$ of G is the minimum value of $m - v(G) - e(G)$ such that an $[m]$-vertex-magic injection of G exists.

There is a significant difference between vertex-magic and edge-magic injections. An edge-magic injection of G corresponds to an edge-magic total labeling of some $G \cup tK_1$. One cannot add an arbitrary number of disjoint vertices in the vertex-magic case, because (Theorem 3.1) a vertex-magic graph cannot contain two isolates, and there seems to be no equivalent notion.

We saw in Theorem 2.46 that every graph has an edge-magic injection. From Theorem 3.1, it is clear that a graph with isolated edge or two isolated vertices could not have a vertex-magic injection. However, all other graphs contain them:

Theorem 3.32 *If a graph has no component K_2 and no two components K_1, then it has a vertex-magic injection.*

Proof. Let G be a graph with no component K_2 and no isolate. We prove that both G and $G \cup K_1$ have vertex-magic injections.

As usual, say G has v vertices and e edges. Write $h = 2^e + 2^{e-1}$. To define a labeling λ, assign the values $1, 2, 2^2, \ldots, 2^{e-1}$ as labels of the edges of G in any order. Then, for each vertex x, define

$$\lambda(x) = h - \sum_{y \sim x} \lambda(xy).$$

This λ clearly has the constant weight property with magic constant h. All the edge labels are distinct, by construction. If x_1 and x_2 are two vertices, then their sets of neighboring edges $\{x_1 y \mid y \sim x_1\}$ and $\{x_2 y \mid y \sim x_2\}$ must be different (these sets are the same only if x_1 and x_2 are both isolates, so that the sets are empty, or $x_1 x_2$ forms a component K_2), so $\sum_{y \sim x_1} \lambda(x_1 y)$ and $\sum_{y \sim x_2} \lambda(x_2 y)$ are sums of different sets of powers of 2, and are unequal. Thus $\lambda(x_1) \neq \lambda(x_2)$, and all the vertex labels are distinct. The largest possible value for the sum of edges neighboring a vertex is $\sum_{i=0}^{e-1} 2^i$, so the smallest possible vertex label is

$$h - \sum_{i=0}^{e-1} 2^i = (2^e + 2^{e-1}) - (2^e - 1) = 2^{e-1} + 1,$$

which is greater than the greatest edge label. So all the labels are different and λ is a vertex-magic injection.

No label could be greater than $h - 1$, so h does not occur as a label. So λ can be extended to a vertex-magic injection of $G \cup K_1$ by assigning label h to the isolate. \square

The preceding proof used the fact that if S is any set of distinct powers of 2, then different subsets of S have different sums. This is of course true if 2 is replaced

by any positive integer n (although it is trivial when $n = 1$, and S has only one member), and is the reason that positional notation works. More generally, we could have used for edge labels any set of positive integers for which all subset sums are distinct (called a *DSS set*). If $\max(S)$ denotes the largest member of the set S, one might hope to find an e-element DSS set S with $\max(S)$ significantly smaller than 2^{e-1}. However, this will not happen. Erdös and Moser [26] proved that if $f(n)$ is the smallest value of $\max(S)$ for any n-element DSS set S, then $f(n) \geq 2^{n-2}n^{-\frac{1}{2}}$. A good survey of the DSS set problem is [18].

4

Totally Magic Labelings

4.1 Basic Ideas

4.1.1 Definitions

In this chapter we investigate the question: for a graph G does there exist a total labeling λ that is both edge-magic and vertex-magic? As we said, such a λ is called a *totally magic labeling* and G is a *totally magic graph*. The constants h and k are the magic constant and magic sum, respectively. We do not require that $h = k$.

We shall see that totally magic graphs are very rare. For this reason, totally magic injections have also been studied. Recall that a graph possessing such a mapping is called a TMI graph.

Most of the material in this chapter comes from [28] and [74], and in the later sections from [72] and [97].

4.1.2 Examples

One quickly constructs three small examples of connected totally magic graphs. An obvious trivial example is the single vertex graph K_1. There are four totally magic labelings of the 3-vertex cycle C_3; if the set of vertex-labels is denoted by S_v, then the labelings have

A.M. Marr and W.D. Wallis, *Magic Graphs*, DOI 10.1007/978-0-8176-8391-7_4, 111
© Springer Science+Business Media New York 2013

$$h = 9, k = 12, S_v = \{4,5,6\}$$
$$h = 10, k = 11, S_v = \{2,4,6\}$$
$$h = 11, k = 10, S_v = \{1,3,5\}$$
$$h = 12, k = 9, S_v = \{1,2,3\}.$$

(There is an obvious duality here.) The three-vertex path P_3 has two labelings. Writing the labels in sequence vertex-edge-vertex-edge-vertex, they are

$$h = 6, k = 9, \text{ labels } 4, 2, 3, 1, 5,$$
$$h = 7, k = 8, \text{ labels } 3, 4, 1, 2, 5.$$

Among the disconnected graphs, only one small example is known: there is exactly one totally magic labeling of $K_1 \cup P_3$, constructed from the first of the P_3-labelings listed above by mapping the isolated vertex to 6.

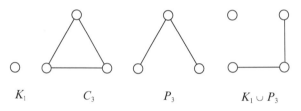

$$K_1 \qquad\qquad C_3 \qquad\qquad P_3 \qquad\qquad K_1 \cup P_3$$

Fig. 4.1. Known totally magic graphs

4.2 Isolates and Stars

We saw in Theorem 3.1 that a graph with two isolated vertices or an isolated edge cannot be vertex-magic. *A fortiori* we have

Lemma 4.1 *No totally magic graph has two isolated vertices or an isolated edge.*

Moreover, if $K_1 \cup G$ is totally magic, the isolated vertex must necessarily receive the largest possible label, so the remaining labels form a totally magic labeling of G. We have:

Lemma 4.2 *If a graph with an isolated vertex is totally magic, then the graph G resulting from the deletion of the isolate has a totally magic labeling with magic constant $|V(G)| + |E(G)| + 1$.*

The labeling of $K_1 \cup P_3$ given above is as described in this lemma.

It was shown in Theorem 3.8 that $K_{m,n}$ is never vertex-magic when $|m - n| > 1$. So no star larger than $K_{1,2}$ is vertex-magic. From this follows:

Lemma 4.3 *No star larger than $K_{1,2}$ is totally magic.*

Theorem 4.4 *Suppose the totally magic graph G has a leaf x. Then the component of G containing x is a star.*

Proof. Suppose λ is a totally magic labeling on G, with magic constant h and magic sum k, and suppose x is a leaf with neighbor y. By the vertex-magic property, $\lambda(x) + \lambda(xy) = h$, and by the edge-magic property, $\lambda(x) + \lambda(xy) + \lambda(y) = k$. So $\lambda(y) = k - h$.

By Lemma 4.1, y has a neighbor, z say. Then $k = \lambda(y) + \lambda(yz) + \lambda(z) = \lambda(yz) + \lambda(z) + k - h$, so $\lambda(yz) = h - \lambda(z)$. So $\mathrm{wt}(z) \geq \lambda(z) + \lambda(yz) = h$, with equality only if z has degree 1. So every vertex adjacent to y has degree 1, and the component of G containing y is a star with center y. □

From Lemma 4.3 and Theorem 4.4, we conclude

Corollary 4.4.1 *The only connected totally magic graph containing a vertex of degree 1 is P_3.*

Every nontrivial tree has at least two vertices of degree 1, so:

Corollary 4.4.2 *The only totally magic trees are K_1 and P_3.*

A totally magic graph cannot have two stars as components, because their centers would each receive label $k - h$. It follows that the components of a totally magic graph can include at most one K_1 and at most one star, and all other components have minimum degree at least 2, and consequently have as many edges as vertices.

Corollary 4.4.3 *The only totally magic proper forest is $K_1 \cup P_3$.*

Theorem 4.5 *The only totally magic graphs with a component K_1 are $K_1 \cup P_3$ and K_1 itself.*

Proof. Suppose $K_1 \cup G$ is a totally magic graph; K_1 has vertex x, and G has v vertices and e edges, as usual. G may have a star as a component, but any other component has minimum degree at least 2, so $e \geq v - 1$. From Lemma 4.2,

$$h = \lambda(x) = v + e + 1,$$

and G is totally magic with magic constant h. Now from (3.3),

$$v(v + e + 1) = \tfrac{1}{2}(v + e)(v + e + 1) + \sum_{y \in E} \lambda(y).$$

So

$$\sum_{y \in E} \lambda(y) = \tfrac{1}{2}(v - e)(v + e + 1).$$

But $\sum \lambda(y) \geq \tfrac{1}{2}e(e + 1)$, so

$$e(e + 1) \leq (v - e)(v + e + 1),$$

and

$$2e(e + 1) \leq v(v + 1).$$

Clearly $e < v$. So $e = v - 1$, whence

$$2(v^2 - v) \leq v^2 + v,$$

and $v \leq 3$. The only possibility is $G = P_3$. $\qquad\square$

Exercise 4.1 *Suppose $K_{1,n} \cup G$ is totally magic, where G has v vertices and e edges. By considering vertex weights in the star, prove that*

$$v + e \geq \tfrac{1}{2}(n^2 - n - 4).$$

4.3 Forbidden Configurations

Theorem 4.6 *If a totally magic graph G contains two adjacent vertices of degree 2, then the component containing them is a cycle of length 3.*

Proof. Suppose G contains a pair of adjacent vertices b and c, each having degree 2, and suppose λ is a totally magic labeling of G with magic constant h and magic sum k.

First, assume G contains a path $\{a, b, c, d\}$, where a and d are distinct vertices. From $h = \mathrm{wt}(b) = \mathrm{wt}(c)$ it follows that

$$\lambda(ab) + \lambda(b) + \lambda(bc) = \lambda(bc) + \lambda(c) + \lambda(cd),$$

from which

$$\lambda(cd) = \lambda(ab) + \lambda(b) - \lambda(c) \qquad\qquad (4.1)$$

while the edge magic property yields

$$k = \lambda(a) + \lambda(ab) + \lambda(b) = \lambda(b) + \lambda(bc) + \lambda(c) = \lambda(c) + \lambda(cd) + \lambda(d); \quad (4.2)$$

the third equality in (4.2) implies

$$\lambda(cd) = \lambda(b) + \lambda(bc) - \lambda(d) \quad (4.3)$$

while the second gives

$$\lambda(a) + \lambda(ab) = \lambda(bc) + \lambda(c). \quad (4.4)$$

But (4.1) and (4.3) give

$$\lambda(ab) - \lambda(c) = \lambda(bc) - \lambda(d)$$

so

$$\lambda(d) = \lambda(bc) + \lambda(c) - \lambda(ab) \quad (4.5)$$

and (4.4) and (4.5) together imply $\lambda(a) = \lambda(d)$, a contradiction.

Now assume b and c have a common neighbor, a say, and suppose a has some other neighbor, z. From (4.2) we see

$$\lambda(ab) = k - \lambda(a) - \lambda(b),$$

and similarly for bc and ca, so

$$\begin{aligned} \mathrm{wt}(b) &= \lambda(ab) + \lambda(b) + \lambda(bc) \\ &= 2k - \lambda(a) - \lambda(b) - \lambda(c), \end{aligned}$$

whereas

$$\begin{aligned} \mathrm{wt}(a) &= \lambda(ca) + \lambda(a) + \lambda(ab) + \lambda(az) + \cdots \\ &\geq 2k - \lambda(a) - \lambda(b) - \lambda(c) + \lambda(az) \\ &> \mathrm{wt}(a), \end{aligned}$$

as $\lambda(az) > 0$. $\qquad\square$

The configurations forbidden by Theorem 4.6 are shown in Fig. 4.2. In this and subsequent figures, small lines attached to a vertex indicate that there may or may not be further edges incident with it.

Corollary 4.6.1 *No totally magic graph contains as a component a path other than P_3 or a cycle other than K_3.*

(Lemma 4.1 must be invoked to rule out P_2.) In particular,

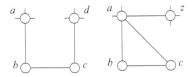

Fig. 4.2. Configurations forbidden by Theorem 4.6

Corollary 4.6.2 *The only totally magic cycle is K_3.*

Theorem 4.7 *Suppose G contains two vertices, x_1 and x_2, that are each adjacent to precisely the same set $\{y_1, y_2, \ldots, y_d\}$ of other vertices. (It is not specified whether x_1 and x_2 are adjacent.) If $d > 1$, then G is not totally magic.*

Proof. Suppose λ is a totally magic labeling of G with magic constant and sum h and k. For convenience, define $\lambda(x_1x_2) = 0$ if x_1 is not adjacent to x_2. Then

$$h = \lambda(x_i) + \sum_{j=1}^{d} \lambda(x_i y_j) + \lambda(x_1 x_2), \ i = 1, 2,$$

and

$$k = \lambda(x_i) + \lambda(x_i y_j) + \lambda(y_j), \ i = 1, 2, \ 1 \leq j \leq d.$$

So

$$dk = d\lambda(x_i) + \sum_{j=1}^{d} \lambda(x_i y_j) + \sum_{j=1}^{d} \lambda(y_j), \ i = 1, 2,$$

$$= (d-1)\lambda(x_i) + (h - \lambda(x_1 x_2)) + \sum_{j=1}^{d} \lambda(y_j), \ i = 1, 2,$$

so

$$(d-1)\lambda(x_1) = (d-1)\lambda(x_2),$$

a contradiction unless $d = 1$. □

(The forbidden configurations appear in Fig. 4.3).

Corollary 4.7.1 *The only totally magic complete graphs are K_1 and K_3. The only totally magic complete bipartite graph is $K_{1,2}$.*

Theorem 4.8 *Suppose G contains two vertices, x and y, with a common neighbor. If x and y are inadjacent and each have degree 2, or are adjacent and each have degree 3, then G is not totally magic.*

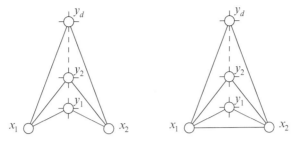

Fig. 4.3. Configurations forbidden by Theorem 4.7

Proof. Denote the common neighbor by z, the other neighbor of x by x_1, and the other neighbor of y by y_1. (Possibly $x_1 = y_1$.) Suppose λ is a totally magic labeling of G with magic constant h and magic sum k; if x is not adjacent to y in G then define $\lambda(xy) = 0$.

The weight of x is

$$\begin{aligned}
\text{wt}(x) &= \lambda(x) + \lambda(xx_1) + \lambda(xz) + \lambda(xy) \\
&= \lambda(x) + (k - \lambda(x) - \lambda(x_1)) + (k - \lambda(x) - \lambda(z)) + \lambda(xy) \\
&= 2k - (\lambda(x) + \lambda(x_1) + \lambda(z)) + \lambda(xy)
\end{aligned}$$

and similarly

$$\text{wt}(y) = 2k - (\lambda(y) + \lambda(y_1) + \lambda(z)) + \lambda(xy).$$

From the vertex-magic property, $\text{wt}(x) = \text{wt}(y)$, so $\lambda(x) + \lambda(x_1) = \lambda(y) + \lambda(y_1)$. So $\lambda(xx_1) = k - \lambda(x) - \lambda(x_1) = k - \lambda(y) - \lambda(y_1) = \lambda(yy_1)$. But this contradicts the edge-magic property. □

(The forbidden configurations appear in Fig. 4.4).

(The case where x and y are adjacent could be rephrased, "a totally magic graph G cannot contain a triangle with two vertices of degree 3.")

Theorem 4.9 *Suppose the totally magic or TMI graph G contains a triangle. Then the sum of the labels of all edges outside the triangle and incident with any one vertex of the triangle is the same, whichever vertex is chosen.*

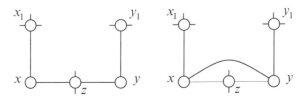

Fig. 4.4. Configurations forbidden by Theorem 4.8

Proof. Suppose the triangle is xyz. Write X for the sum of the labels of all edges other than xy and xz that are adjacent to x; define Y and Z similarly. Then

$$
\begin{aligned}
h = \mathrm{wt}(x) &= \lambda(x) + \lambda(xy) + \lambda(xz) + X = 2k - \lambda(x) - \lambda(y) - \lambda(z) + X \\
= \mathrm{wt}(y) &= \lambda(y) + \lambda(xy) + \lambda(yz) + Y = 2k - \lambda(x) - \lambda(y) - \lambda(z) + Y \\
= \mathrm{wt}(z) &= \lambda(z) + \lambda(xz) + \lambda(yz) + Z = 2k - \lambda(x) - \lambda(y) - \lambda(z) + Z
\end{aligned}
$$

so $X = Y = Z$. □

Corollary 4.9.1 *If the totally magic graph G contains a triangle with one vertex of degree 2, then the triangle is a component of G.*

Theorem 4.10 *Suppose vertices x, y, z, t form a 4-cycle in the totally magic or TMI graph G. Denote by $\tau(x)$ the sum of weights of all the edges touching x, other than the edges in the cycle. Then $\tau(x), \tau(y), \tau(z), \tau(t)$ are all different.*

Proof. We have $\tau(x) = \mathrm{wt}(x) - \lambda(x) - \lambda(xy) - \lambda(xt)$; $\tau(y), \tau(z)$ and $\tau(t)$ are defined similarly. We have

$h = \lambda(xy) + \lambda(x) + \lambda(xt) + \tau(x)$	$k = \lambda(x) + \lambda(xy) + \lambda(y)$
$h = \lambda(yz) + \lambda(y) + \lambda(yx) + \tau(y)$	$k = \lambda(y) + \lambda(yz) + \lambda(z)$
$h = \lambda(zt) + \lambda(z) + \lambda(zy) + \tau(z)$	$k = \lambda(z) + \lambda(zt) + \lambda(t)$
$h = \lambda(tx) + \lambda(t) + \lambda(tz) + \tau(t)$	$k = \lambda(t) + \lambda(tx) + \lambda(x)$

From the first and second "h" equations, we get

$$
\tau(x) - \tau(y) = (\lambda(y) + \lambda(yz)) - (\lambda(tx) + \lambda(x)).
$$

The second and fourth "k" equations give $(\lambda(y)+\lambda(yz)) = k-\lambda(z)$ and $(\lambda(x)+\lambda(tx)) = k - \lambda(t)$, respectively. So

$$
\tau(x) - \tau(y) = k - \lambda(z) - (k - \lambda(t)) = \lambda(t) - \lambda(z)
$$

which cannot be zero. So $\tau(x) \neq \tau(y)$. $\tau(y) \neq \tau(z)$, $\tau(z) \neq \tau(t)$ and $\tau(t) \neq \tau(x)$ are proven similarly. To show that $\tau(x) \neq \tau(z)$, we can use the first and third "h" equations to show

$$
\tau(x) - \tau(z) = (\lambda(yz) + \lambda(z) + \lambda(zt)) - (\lambda(tx) + \lambda(x) + \lambda(xy)).
$$

Now the first and second "k" equations yield $\lambda(yz) + \lambda(z) = \lambda(x) + \lambda(xy)$, so $\tau(x) - \tau(z) = \lambda(zt) - \lambda(tx)$, which again cannot be zero, so $\tau(x) \neq \tau(z)$, and similarly $\tau(y) \neq \tau(t)$. □

In the above proof, if x has degree 2 in G, $\tau(x)$ must be zero. So

Corollary 4.10.1 *If a graph G contains a 4-cycle in which two or more vertices are of degree 2 in G, then G does not have a totally magic labeling or injection.*

Observe that Theorems 4.6–4.10, and their corollaries are essentially *forbidden configuration* theorems. If a graph G is in violation of one of them, then not only is G not totally magic, but G cannot be a TMI graph or a component or union of components in any totally magic graph or TMI graph.

Exercise 4.2 *Suppose xy is any edge of a totally magic graph whose labeling λ has magic constant and sum h and k. Prove that*

$$\lambda(xy) + \sum_{\substack{z \sim x \\ z \neq y}} \lambda(xz) + \sum_{\substack{t \sim y \\ t \neq x}} \lambda(xz) = 2h - k.$$

Exercise 4.3 (The Star Theorem.) *Suppose x is any vertex of a totally magic graph and x_1, x_2, \ldots, x_n are the vertices adjacent to it. Denote by τ_i the sum of weights of all the edges touching x_i, other than xx_i. Prove that $\tau_1, \tau_2, \ldots, \tau_n$ are all equal.*

4.4 Unions of Triangles

In this section we construct two infinite families of totally magic graphs, both based on triangles. We shall show that the union of an odd number of triangles is always totally magic, and so is the graph constructed by deleting one edge from such a union.

Suppose λ is a totally magic labeling of nK_3, with magic constant h and magic sum k, and suppose n is odd. Suppose the vertices of one of the triangles are labeled x, y, z; suppose the edge opposite the vertex labeled x receives the label X, and so on. Then

$$h = x + Y + Z = X + y + Z = X + Y + z, \tag{4.6}$$

$$k = X + y + z = x + Y + z = x + y + Z. \tag{4.7}$$

If we write $s = x + y + z$ and $S = X + Y + Z$, then (4.6) and (4.7) yield

$$\begin{aligned} 3h &= 2S + s, \\ 3k &= S + 2s, \\ h + k &= S + s. \end{aligned}$$

So

$$3h - 3k = S - s,$$

whence

$$S = 2h - k,$$
$$s = 2k - h.$$

Finally, from $X + Y + Z = S$ and $x + Y + Z = h$ we obtain

$$X - x = S - h = h - k \tag{4.8}$$

for every choice of x, that is the difference between an edge label and the vertex label is a constant, $h - k$.

Let us write d for $h - k$. It is clear that h and k determine d. On the other hand, if d is known, then h and k are determined. For each triangle, $S + s = h + k$, so summing over all triangles we get $n(h + k)$. But this is the sum of all vertex and edge labels in nK_3, so it equals $\frac{1}{2}6n(6n + 1)$, and

$$h + k = 3(6n + 1). \tag{4.9}$$

Therefore

$$h = 9n + \tfrac{1}{2}(3 + d) \text{ and } k = 9n + \tfrac{1}{2}(3 - d). \tag{4.10}$$

Without loss of generality, let us assume $k < h$, that is $d = h - k > 0$. ($k = h$ is impossible, as it would imply $x = X$, and if $k > h$ then we can obtain a *dual* labeling with $k < h$ by interchanging the labels on each vertex and its opposite edge.) From (4.8), every edge label is at least $d+1$, and $1, 2, \ldots, d$ are all vertex labels. The edges opposite those vertices receive labels $d+1, d+2, \ldots, 2d$. Proceeding in this way, we see that the set of vertex labels is

$$T_d(n) = \{1, 2, \ldots, d, 2d + 1, 2d + 2, \ldots, 3d, 4d + 1, \ldots, 6n - d\}$$

and the edge labels are

$$\{d + 1, d + 2, \ldots, 2d, 3d + 1, 3d + 2, \ldots, 4d, 5d + 1, \ldots, 6n\}.$$

So $6n$ is a multiple of $2d$. We have

Theorem 4.11 *If there is a totally magic labeling of nK_3 where n is odd, then the magic constant and sum have the values*

$$h = 9n + \tfrac{1}{2}(3 + d) \text{ and } k = 9n + \tfrac{1}{2}(3 - d), \tag{4.10}$$

for some divisor d of $3n$.

(The negative divisors d correspond to the labelings that are duals of those with positive d.)

Suppose d is a specified positive divisor of $3n$. Let $3n = ad$. Then the sum of the elements of $T_d(n)$ is

$$ns = \sum_{i=1}^{d}\sum_{j=0}^{a-1}(2jd+i)$$

$$= \sum_{i=1}^{d}(a(a-1)d+ai)$$

$$= a(a-1)d^2 + \tfrac{1}{2}ad(d+1),$$

$$s = 3(a-1)d + \tfrac{3}{2}(d+1). \tag{4.11}$$

So the triples of vertex labels of triangles must form a partition of $T_d(n)$ into n triples, each with sum (4.11). Conversely, any such partition will define a totally magic labeling with the constants (4.10).

Lemma 4.12 *If n is an odd positive integer and d is any divisor of $3n$, then there exists a partition of*

$$T_d(n) = \{1, 2, \ldots, d, 2d+1, 2d+2, \ldots, 3d, 4d+1, \ldots,$$
$$(2a-2)d+1, (2a-2)d+2, \ldots, (2a-1)d\}$$

into n triples, each with sum (4.11), where $a = 3n/d$.

Proof. We use two families of $3 \times m$ arrays, m odd, which are denoted A_m and B_m, and are defined as follows. A_m has j-th column $(j, m+2-2j, \tfrac{1}{2}(m-1)+j)$, where entries are reduced modulo m, if necessary, so that they lie in the range $1, 2, \ldots, m$. This array was used by Kotzig [52]. Each row is a permutation of $\{1, 2, \ldots, m\}$, and each column sums to $\tfrac{1}{2}(3m+3)$. B_m is constructed from a copy of A_m by adding m to every entry in the second row and $2m$ to every entry in the third row, so each column has sum $\tfrac{1}{2}(9m+3)$.

First, suppose d is a multiple of 3. Form a master array M by subtracting 1 from each entry of A_a and multiplying by $2d$. From each column of M we construct $d/3$ triples by adding each of the columns of $B_{d/3}$. It is easy to check that the $a(d/3) = n$ triples form a partition of $T_d(n)$. All the triples have the same sum, because both component arrays have constant column sum.

If d is not a multiple of 3, then a must be. In that case the master array is formed from $A_{a/3}$, again subtracting 1 and multiplying by $2d$. The columns of B_d are added. \square

Table 4.1 provides an illustration of the two constructions.

Theorem 4.13 *Suppose n is odd. There is a totally magic labeling of nK_3 with magic constant h and magic sum k if and only if $h = 9n + \tfrac{1}{2}(3+d)$ and $k = 9n + \tfrac{1}{2}(3-d)$, where d is a divisor of $3n$.*

$$M = \begin{bmatrix} 0 & 18 & 36 & 54 & 72 \\ 72 & 36 & 0 & 54 & 18 \\ 36 & 54 & 72 & 0 & 18 \end{bmatrix} \quad B_{d/3} = \begin{bmatrix} 1 & 2 & 3 \\ 6 & 4 & 5 \\ 8 & 9 & 7 \end{bmatrix}$$

Partition

1	19	37	55	73	2	20	38	56	74	3	21	39	57	75
78	42	6	60	24	76	40	4	58	22	77	41	5	59	23
44	62	80	8	26	45	63	81	9	27	43	61	79	7	25

Example for $a = 5, d = 9, n = 15$, column sum 123.

$$M = \begin{bmatrix} 0 & 10 & 20 \\ 50 & 30 & 40 \\ 70 & 80 & 60 \end{bmatrix} \quad B_d = \begin{bmatrix} 1 & 2 & 3 & 4 & 5 \\ 5 & 3 & 1 & 4 & 2 \\ 3 & 4 & 5 & 1 & 2 \end{bmatrix}$$

Partition

1	2	3	4	5	11	12	13	14	15	21	22	23	24	25
55	53	51	54	52	35	33	31	34	32	45	43	41	44	42
73	74	75	71	72	83	84	85	81	82	63	64	65	61	62

Example for $a = 9, d = 5, n = 15$, column sum 129.

Table 4.1. Partitions for labeling $15K_3$

This Theorem not only proves the existence of totally magic labelings of odd unions of triangles, but it also determines the spectrum of realizable magic constants for such labelings. However, there may be many non-isomorphic labelings that give the same constants. For example, there are exactly eight totally magic labelings of $3K_3$. They arise in dual pairs (the dual is derived by exchanging the label of each vertex with that of its opposite edge). Of the four pairs, two come from the construction of Theorem 4.13 and two do not.

In the cases where d is negative, 1 is always an edge label. Delete this edge, and subtract 1 from every label. The resulting labeling is totally magic, with magic constant and sum found by subtracting 3 from the original constants. We have another infinite family:

Corollary 4.13.1 *The graph $P_3 \cup nK_3$ is totally magic when n is even.*

Exercise 4.4 *Prove that every totally magic labeling of $P_3 \cup nK_3$ comes from a labeling of $(n+1)K_3$ with one edge labeled 1, in the manner described above.*

Exercise 4.5 *What are the possible magic constants for totally magic labelings of $P_3 \cup nK_3$?*

In each dual pair of labelings of $3K_3$, one member has 1 as an edge label, so (from Exercise 4.4) there are exactly four totally magic labelings of $P_3 \cup 2K_3$.

Theorem 4.14 *There is no totally magic labeling of nK_3, the disjoint union of n triangles, when n is even.*

Proof. Let us write $n = 2m$. Again we assume that $k < h$, write $d = h - k$, and consider partitioning the set $T_d(2m)$ and its complement into triples.

First, suppose $h - k = 1$. We have to partition the edge labels $2, 4, \ldots, 12m$ into $2m$ sets of size 3 such that the sum of the labels in each set is constant. If such a partition exists, halving it provides a partition of $1, 2, \ldots, 6m$ into $2m$ sets of size 3 such that the sum of the labels in each set is a constant. But if that common sum is S, then $2mS = \sum_{i=1}^{2m} i = m(2m+1)$, and $S = \frac{1}{2}(2m+1)$, which is not an integer. So $h - k = 1$ is impossible.

So we assume $h - k \geq 3$. Write $h - k = 2c + 1$. From (4.10) we obtain $h = 18m + c + 2$ and $k = 18m - c + 1$. So $2k - h = 18m - 3c$, and $2k - h \equiv 18m - 3c \equiv -3c \equiv c + 2 \pmod{4c + 2}$, since $2c + 1$ divides $3m$.

Each vertex label must be congruent to one of $1, 2, \ldots, h - k \pmod{2(h - k)}$, that is $1, 2, \ldots, 2c + 1 \pmod{4c + 2}$. The sum of the three vertex labels in any triangle will be congruent to $2k - h$. Now, $2k - h \equiv c + 2 \pmod{4c + 2}$. Since $0 < c < h - k, c + 1$ must be a vertex label. But this is impossible, since there do not exist two elements $d, e \in \{1, 2, \ldots, 2c + 1\} \pmod{4c + 2}$ such that $c + 1 + d + e \equiv c + 2 \pmod{4c + 2}$.

Hence $2mK_3$ is not totally magic for any positive integer m. □

Corollary 4.14.1 *The graph $P_3 \cup nK_3$ is not totally magic when n is odd.*

Proof. Suppose there is a totally magic labeling of $P_3 \cup nK_3$ with vertex and edge constants h and k. The center vertex of P_3 must receive label $k - h$. Suppose the other vertices of P_3 receive a and b. Then the two edges must be labeled $h - a$ and $h - b$. So the center vertex has weight $(k - h) + (h - a) + (h - b) = k - h - a - b$. Therefore $k - h - a - b = h$, and $a + b = k$. One can construct a totally magic labeling of $(n + 1)K_3$ with vertex and edge constants $h + 3$ and $k + 3$ by adding 1 to every label on $P_3 \cup nK_3$ and joining the endpoints of P_3 with an edge labeled 1. When n is odd, this contradicts Theorem 4.14. □

In the first edition of this text, Research Problem 4.1 asked if the graph $K_{1,m} \cup nK_3$ is ever totally magic. As Corollary 4.13.1 shows, $K_{1,2} \cup nK_3$ is

totally magic when n is even. And, Calhoun *et al.* showed that this is the only case in which $K_{1,m} \cup nK_3$ has a totally magic labeling [20]. Corollary 4.14.1 shows $K_{1,2} \cup nK_3$ is not totally magic when n is odd. The following theorem excludes the remaining cases:

Theorem 4.15 *For any $n \geq 0$ and $m \geq 1$, suppose G is a totally magic graph isomorphic to $K_{1,m} \cup nK_3$. Then $m = 2$.*

Proof. Let λ be a totally magic labeling for G with vertex sum h and edge sum k. Let $d = k - h$. Let c be the central vertex of the star and b_1, b_2, \ldots, b_m be the other vertices of the star. Then since $\lambda(b_1) + \lambda(b_1 c) + \lambda(c) = k$ and $\lambda(b_1) + \lambda(b_1 c) = h$, we subtract to get that $\lambda(c) = d$. As there are $3n + m + 1$ vertices and $3n + m$ edges, the maximum label

$$M = 6n + 2m + 1. \tag{4.12}$$

The sum of the labels is $M(M + 1)/2$. However, the sum of the labels on each K_3 is $h + k$, and for each i, $\lambda(b_i) + \lambda(b_i c) = h$, so we also get that the sum of the labels is $d + mh + n(h + k) = d + mh + n(2h + d) = (2n + m)h + (n + 1)d$. Setting these two expressions equal we get

$$h = \frac{M(M + 1)/2 - (n + 1)d}{2n + m}. \tag{4.13}$$

Also, $\sum \lambda(b_i) \leq M + (M - 1) + \cdots + (M - m + 1) = mM - m(m - 1)/2 = m(M - (m - 1)/2)$. And since $\lambda(b_i c) = h - \lambda(b_i)$ and $\sum \lambda(b_i c) = h - d$ we get $h - d \geq mh - m(M - (m - 1)/2)$ and hence

$$h(m - 1) \leq m(M - (m - 1)/2) - d. \tag{4.14}$$

Using equations (4.13) and (4.14) we get

$$\frac{(m - 1)(M(M + 1)/2 - (n + 1)d)}{2n + m} \leq m(M - (m - 1)/2) - d$$

Now using equation (4.12) we get

$$(m - 1)((6n + 2m + 1)(3n + m + 1) - (n + 1)d)$$
$$\leq (2n + m)(m(6n + 3m/2 + 3/2) - d).$$

Expanding, rearranging and factoring gives us:

$$(m - 3)(6n^2 + 3(m + 1)n + m(m + 2)/2) + m - 1 + d \leq (m - 3)nd. \tag{4.15}$$

If $m \geq 3$, then $(m - 3)m(m + 2)/2 + m - 1 + d > 0$ and hence equation (4.15) becomes

$$(m - 3)(6n^2 + 3(m + 1)n) < (m - 3)md. \tag{4.16}$$

Hence, if $m = 1$, then $\lambda(b_1) = d = \lambda(c)$, so G could not be totally magic. If $m = 3$ or $m > 3$ and $n = 0$, then equation (4.16) implies $0 < 0$. If $m > 3$ and $n > 0$, then $M = 6n + 2m + 1 < 6n + 3(m + 1) < d$. But, since $\lambda(c) = d$ and M is the largest label $M \geq d$.

Hence, all that remains is that $m = 2$. □

4.5 Small Graphs

A complete search for small totally magic graphs is described in [28]. The result is that no examples with ten or fewer vertices exist, other than the four graphs given in Sect. 4.1.2 and the two nine-vertex graphs constructed in Theorem 4.13 and Corollary 4.13.1.

The complete search was carried out in two stages. First, using *nauty* [68], lists were prepared of all connected graphs (up to ten vertices) not ruled out by Theorems 4.6–4.8 and Corollary 4.9.1. Second, the survivors were tested exhaustively. There were four survivors with six or fewer vertices (K_1, K_3, and P_3 and the graph shown in Fig. 4.5), 42 with seven, 1,070 with eight, 61,575 with nine and 4,579,637 with 10. The graph of Fig. 4.5 was shown to have no totally magic labeling or injection using Theorem 4.9 (see Exercise 4.6); probably many of the other survivors could also be eliminated by ad hoc applications of that Theorem.

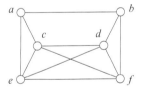

Fig. 4.5. A graph to be eliminated using Theorem 4.9

Exhaustive testing is very time-consuming. However, a shortcut is available. If a totally magic labeling exists, it must retain the magic properties after reduction modulo 2. So we tested all $\mod 2$ possibilities. Label 1 or 0 is assigned to each vertex and to the constant k. Then every edge-label can be calculated. Next, one can check whether all the vertex weights are congruent $(\mod 2)$. Moreover, the total number of vertices and edges labeled 1 must either equal the number labeled 0 or exceed that number by 1. This process is quite fast (e.g., only 2^9 cases need to be examined in the eight-vertex case), and eliminated over 25% of graphs. Then one can sieve the remaining graphs modulo 3, then modulo 4, and so on. For example, sieving the 1070 eight-vertex graphs mod 2 eliminated 307 graphs, sieving mod 3 eliminated 351 more, and so on: one graph survived after sieving modulo 7, and it was eliminated mod 8.

There were very few disconnected graphs to consider. Except for the graph $K_1 \cup P_3$, the only possibilities are made up of at most one star, copies of K_3, and survivors with more than three vertices. The only cases not already discussed are $K_{1,n} \cup K_3$ for $n = 3, 4, 5, 6$, $K_{1,3} \cup 2K_3$, and the 42 unions of a triangle and a seven-vertex survivor. None of these is totally magic, so there are no further totally magic graphs with ten or fewer vertices.

A further investigation using a variant of simulated annealing has been carried out. This procedure quickly finds the graphs we have described, but has so far found no other examples. This might suggest that we have found all totally magic graphs. However, for larger numbers of vertices (more than 20, say), it appears that the search gets "nearer" to satisfaction (no, we do not wish to clarify this vague description!), so perhaps there are large totally magic graphs yet to be discovered.

Exercise 4.6 *Show that the graph of Fig. 4.5 has no totally magic labeling or injection.*

Research Problem 4.1 *Prove or disprove: no graph of the form $K_{1,n} \cup G$ is totally magic, when $n \geq 3$.*

Research Problem 4.2 *Are there any further totally magic graphs?*

4.6 Totally Magic Injections

Much of the material in this section is taken from [72].

As we said in Sect. 1.4, *totally magic injection* is an injection that is both edge-magic and vertex-magic. We define an $[m]$-*totally magic injection of* G to be a totally magic injection of G in which the largest label is m, and again call m the *size* of the injection. If G has a totally magic injection, G is called a *TMI graph*; the *total deficiency* $def_t(G)$ of G is the minimum value of $m - v(G) - e(G)$, such that there is an $[m]$-totally magic injection of G, and the corresponding minimal value of m is denoted $m_t(G)$. A labeling achieving this bound is called *minimal*.

Theorems 4.6–4.9 all show that certain classes of graphs have no totally magic injection, so totally magic injections are somewhat rarer than the edge-magic and vertex-magic species. However, some graphs are known to have totally magic injections but are not totally magic.

Theorem 4.16 *The star $K_{1,n}$ has a totally magic injection provided $n > 1$. The total deficiency when $n > 2$ is $def_t(K_{1,n}) = \binom{n+2}{2} - 2n - 3$.*

Proof. Suppose λ is a totally magic injection of a $K_{1,n}$ with center c and leaves x_1, x_2, \ldots, x_n; write a_i for $\lambda(cx_i)$, and let a be the smallest a_i. From the proof of Theorem 4.4, $\lambda(c) = k - h$, so $k = \text{wt}(cx_i) = (k - h) + a_i + \lambda(x_i)$ implies $\lambda(x_i) = h - a_i$. The weight of c is $(k - h) + \sum a_i$, and this will equal h. So h is the sum of $n + 1$ positive integers, whence $h \geq \binom{n+2}{2}$. The largest vertex weight, $h - a$, is clearly the largest value of λ. Now $h - a$ is the sum of n positive integers, $n - 1$ of which are greater than a, so it must equal at least $1 + 3 + 4 + \cdots + (n+1)$, or $\binom{n+2}{2} - 2$. So this is the smallest possible size of a totally magic injection.

This size can be realized when $n > 2$. Take $h = \binom{n+2}{2}, k = h + 1, a_i = i + 1$. Then $\lambda(c) = 1, \lambda(cx_i) = \binom{n+2}{2} - i - 1$. This totally magic injection has deficiency $\binom{n+2}{2} - 2 - v(K_{1,n}) - e(K_{1,n}) = \binom{n+2}{2} - 2n - 3$.

When $n = 2$, this construction gives two equal labels. However $K_{1,2}$ is known to be totally magic. \square

It follows from the discussion in Sect. 4.2 that a graph with two isolates cannot have a totally magic injection. However, graphs with one isolate are more easily handled.

Theorem 4.17 *Suppose G has a totally magic injection λ with magic constant h. If G has no isolated vertex, then $G \cup K_1$ has a totally magic injection of size h.*

Proof. Extend λ to $G \cup K_1$ by labeling the isolate with h. All that is necessary is to check that h is not already in use as a label on G. But if either $\lambda(x) = h$ or $\lambda(xy) = h$, this would imply that $\text{wt}(x) > h$ (there are no isolates), which is impossible. \square

Suppose this method is applied to the labeling of $K_{1,n}$ constructed in Theorem 4.16. It is clear that $\text{wt}(c)$ cannot be smaller than $\binom{n+2}{2}$ in any totally magic injection, so the isolated vertex in $K_{1,n} \cup K_1$ could not receive a smaller label than $\binom{n+2}{2}$. Therefore the application of Theorem 4.17 to a minimal injection of $K_{1,n}$ produces a minimal injection of $K_{1,n} \cup K_1$. But there is no reason why this should be true in general.

Research Problem 4.3 *Find examples in which, when the method of Theorem 4.17 is applied to a minimal totally magic injection of G, the resulting injection of $G \cup K_1$ is not minimal.*

Exercise 4.7 *Prove that the only forests with totally magic injections are K_1, $K_{1,n}$ for $n \geq 2$, and $K_1 \cup K_{1,n}$ for $n \geq 2$. For each forest F, find $m_t(F)$.*

If a disconnected graph is totally magic (or has a totally magic injection), the totally magic labeling induces a totally magic injection on its components. For example, the union of an *even* number of triangles may be constructed from the

union of an odd number by deleting one triangle. So it follows from Theorem 4.13 and Corollary 4.13.1 that $2nK_3$ has a totally magic injection, as does the graph obtained from it by deleting one edge. However, consider the following alternate labeling of nK_3 for n even given in [20]:

Let $x_i = 2i - 1$ be one vertex label in each of the n triangles. Then, proceeding counter clockwise in each triangle label the next two vertices

$$y_i = \begin{cases} 4n + 1 - i & \text{if } i \text{ is even} \\ 3n - i & \text{if } i \text{ is odd} \end{cases}, \quad z_i = \begin{cases} 5n + 1 - i & \text{if } i \text{ is even} \\ 6n + 2 - i & \text{if } i \text{ is odd.} \end{cases}$$

Then, the edge opposite vertex x_i is given the label $X_i = x_i + 1$. Similarly, $Y_i = y_i + 1$ and $Z_i = z_i + 1$. This gives $h = 9n + 3$ and $k = 9n + 2$ and it gives a totally magic injection of nK_3 with largest label $6n + 2$. Hence, using this labeling and Theorem 4.14 we have that the total deficiency of nK_3 for n even is either 1 or 2. In [20], the deficiency is shown to be 2. This solves Research Problem 4.5 given in first edition of this text. However, the proof is somewhat lengthy and is omitted here.

By Theorem 4.15, we know $K_{1,m} \cup nK_3$ is totally magic if and only if $m = 2$ and n is even. The following results gives that $K_{1,m} \cup nK_3$ is a TMI graph for all $n \geq 0$ and $m > 1$.

Lemma 4.18 [20] *For all positive integers n and m, there is a totally magic injection of nK_3 with vertex magic number h and edge magic number k such that*

$$\frac{m(m+1)}{2} \leq 2h - k$$

and every label is greater than $d = k - h$.

Proof. By Theorem 4.13 and the discussion above, we know there is a totally magic injection of nK_3, say with $h = a$ and $k = b$. We get a new totally magic injection by adding c to every vertex label where $c = \max\{d, (m(m + 1)/2 - 2a + b)/3\}$. The result is a totally magic injection with $h = a + 3c$ and $k = b + 3c$. Note that d is unchanged. Also, the last label used in the new labeling is at least $1 + c > d$ and

$$2h - k = 2(a+3c) - (b+3c) \geq 2a - b + 3(m(m+1)/2 - 2a+b)/3 = m(m+1)/2.$$

□

Theorem 4.19 [20] *For any $n \geq 0$ and $m > 1$, $K_{1,m} \cup nK_3$ is a TMI graph.*

Proof. The case when $n = 0$ is handled by Theorem 4.16 and the case when $m = 2$ is handled by the paragraph following Exercise 4.7. Thus, we assume $n \geq 1$

and $m \geq 3$. Consider a total labeling of the triangles that satisfies Lemma 4.18, say with $h = a$ and $k = b$. We may assume $a < b$. Multiply each label by s, where $s = m(m+1)/2 + 1$. This gives us a totally magic injection of the triangles with $h = sa$, $k = sb$, and $d = sb - sa$. Note that by the conditions of Lemma 4.18 each label is greater than d. Now label the central vertex of the star d. Label $m - 1$ of the edges $1, 2, \ldots, m - 1$ and label the corresponding vertices $sa - 1, sa - 2, \ldots, sa - m + 1$. Label the last edge $s(2a - b) - m(m-1)/2$, and label the corresponding vertex $s(b - a) + m(m - 1)/2$. Note that none of these labels are divisible by s, so they have not been used in the labeling of the triangles of the central vertex. Also, by the conditions of Lemma 4.18,

$$s(2a - b) - m(m - 1)/2 \geq sm(m + 1)/2 - m(m - 1)/2 \geq m.$$

Hence the labels on the edges are all distinct. Since $sa + m - 1 < sb + m(m-1)/2$, we also have that $s(2a - b) - m(m - 1)/2 < sa - m + 1$. Clearly,

$$m - 1 < s(b - a) + m(m - 1)/2.$$

Furthermore, by the conditions of Lemma 4.18,

$$s(2a - b) \geq 2a - b \geq m(m + 1)/2 > m(m - 1)/2 + m - 1.$$

So $sa - m + 1 > s(b - a) + m(m - 1)/2$. It follows that all the labels are distinct except possibly $s(2a - b) - m(m - 1)/2$. Then $(m(m - 1)/2 + 1)(3a - 2b) = m(m - 1)$ or equivalently $(1 + \frac{2}{m(m-1)})(3a - 2b) = 2$. Thus $0 < 3a - 2b < 2$. Since $3a - 2b$ is an integer $3a - 2b = 1$. But then $m = 2$, so this case cannot occur. We have shown all the labels are distinct. It is now an easy exercise to verify that the magic equations hold. So we have a totally magic injection of the graph $K_{1,m} \cup nK_3$. \square

Exercise 4.8 *Show the labeling given in Theorem 4.19 is a totally magic injection for $K_{1,m} \cup nK_3$.*

The previous proof shows the existence of a totally magic injection of $K_{1,m} \cup nK_3$ for $n \geq 0$ and $m > 1$, but not attempt has been made to achieve the minimum total deficiency. Hence, the following problem was given in [20].

Research Problem 4.4 *What is the total deficiency of $K_{1,m} \cup nK_3$?*

From the results we have seen so far, the known TMI graphs with seven or fewer vertices are shown in Table 4.2.

	TMGs	Others
1 vertex:	K_1	None
2 vertices:	None	None
3 vertices:	K_3, P_3	None
4 vertices:	$P_3 \cup K_1$	$K_3 \cup K_1, K_{1,3}$
5 vertices:	None	$K_{1,3} \cup K_1, K_{1,4}$
6 vertices:	None	$K_3 \cup K_3, K_3 \cup P_3, K_{1,4} \cup K_1, K_{1,5}$
7 vertices:	None	$K_3 \cup K_3 \cup K_1, K_3 \cup P_3 \cup K_1,$
		$K_{1,5} \cup K_1, K_{1,6}, K_{1,3} \cup K_3$

Table 4.2. Known small TMI graphs

4.7 The Totally Magic Equation Matrix

Suppose λ is a totally magic labeling of a graph G with vertices X_1, X_2, \ldots, X_v and edges Y_1, Y_2, \ldots, Y_e.

$$
\begin{aligned}
\mathrm{wt}(X_i) - h &= 0 \quad \text{for all } i \in \{1, 2, \ldots, v\}, \\
\mathrm{wt}(Y_j) - k &= 0 \quad \text{for all } j \in \{1, 2, \ldots, e\}.
\end{aligned}
\tag{4.17}
$$

Let us write M_G for the matrix of coefficients of this system of equations. M_G is the *totally magic equation matrix* of G. For convenience, we refer to the row corresponding to the weight equation for vertex X_i or edge Y_j as "row X_i" or "row Y_j," respectively. Similarly we refer to columns by the appropriate vertex, edge or magic constant.

The problem of determining whether G is a TMI graph can thus be stated as the problem of finding positive integer solutions to

$$
M_G[x_1, x_2, \ldots, x_v, y_1, \ldots, y_e, h, k]^{\mathrm{T}} = 0
\tag{4.18}
$$

where all the x's and y's are distinct (and then setting $\lambda(X_i) = x_i$, etc.).

We write M_G^- for the matrix derived from M_G by deleting the columns for h and k. We denote the determinant of M_G^- by $|M_G^-|$.

Theorem 4.20 *If β is a totally magic injection on a graph G, and M_G^- is invertible, then each edge label has the form*

$$
\beta(Y_i) = \frac{n_i}{|M_G^-|}(2h - k),
\tag{4.19}
$$

where the n_i are integers.

Proof. Equation (4.18) can be interpreted as the set of equations

$$M_G^-[x_1, x_2, \ldots, x_v, y_1, \ldots, y_e]^{\mathrm{T}} = b,$$

where b is the vector with its first v elements k and its other e elements h.

Given the edge Y_k joining vertices X_i and X_j, consider row Y_k of M_G^-. It contains 1's in columns Y_k, X_i and X_j and 0's elsewhere. Rows X_i and X_j will contain 1's in columns X_i and X_j, respectively, and in the columns of edges incident with the relevant vertex. Subtracting rows X_i and X_j from row Y_k and then negating yields a row with 0 in each vertex column, 1 in each column representing an edge incident with X_i and X_j (or both), and 0 in the other edge columns. The right hand will be $2h - k$. If we repeat for all edge rows, we finish with the last e equations in the form $\sum y_i = 2h - k$.

If we continue Gaussian elimination until the first $v + e$ columns form an identity matrix (possible because M_G^- is invertible), only edge rows will be added to the edge rows, so the right-hand sides of the final e equations will all be multiples of $2h - k$. Cramer's rule tells us that the multiplier must be an integer multiple of the inverse of $|M_G^-|$. □

In the process of proving Theorem 4.20, we in fact showed

Corollary 4.20.1 *If $X_i X_j$ is an edge of a TMI graph, the sum of the weights of all edges incident with X_i or X_j or both is $2h - k$.*

If G is regular, an analogous proof applies to the vertex weights.

Theorem 4.21 *If β is a totally magic injection on a regular graph G of degree d, and M_G^- is invertible, then each vertex label has the form*

$$\beta(X_i) = \frac{m_i}{|M_G^-|}(dk - h), \tag{4.20}$$

where the m_i are integers.

4.8 Survivors on Seven Vertices

The elimination process at the beginning of Sect. 4.5 shows that the only possible TMI graphs with fewer than seven vertices that were not eliminated by the above theorems were those we have already listed in Sect. 4.6. There were 42 connected

graphs on 7 vertices, other than the star, not eliminated by the theorems; as we said, these survivors were eliminated by computer-aided techniques (see [28]). Those techniques did not rule out the possibility that the graphs might have totally magic injections. We examine these 42 graphs. To identify the graphs, we use the labeling in the database of graphs on seven or fewer vertices given in [1].

In all 42 cases, it turned out that M_G^- was invertible. Therefore row reduction of (4.18) leads to a set of equations of the form

$$x_i = \frac{h_i}{|M_G^-|}h + \frac{k_i}{|M_G^-|}k, 1 \le i \le v,$$

$$y_j = \frac{n_j}{|M_G^-|}(2h - k), 1 \le j \le k.$$

For a graph G, we write

$$P_G^h = [h_1, h_2, \ldots, h_v],$$
$$P_G^k = [k_1, k_2, \ldots, k_v] \text{ and}$$
$$N_G = [n_1, n_2, \ldots, n_e].$$

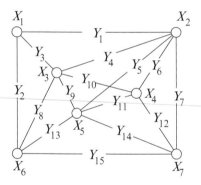

Fig. 4.6. The graph G_{1200}

Figure 4.6 shows G_{1200}, in the notation of [1]. We found that this is a TMI graph. It has $|M_{G_{1200}}^-| = -789$, and satisfies

$$P_{G_{1200}}^h = [359, -7, 64, 169, 97, 47, 161],$$
$$P_{G_{1200}}^k = [-574, -391, -410, -479, -443, -418, -475] \text{ and}$$
$$N_{G_{1200}} = [-176, -203, -195, -12, -45, -81, -77, -39, -64,$$
$$-100, -133, -165, -72, -129, -104].$$

To ensure that all labels are integers, we need $2h - k \equiv 0 \pmod{789}$. If $2h - k \le 0$ the edge-labels will be non-positive, so we try the minimum feasible value,

$2h - k = 789$. Then $h = k = 789$ is an obvious solution, and others can be found by adding t to each vertex label, giving $h = 789 + t$ and $k = 789 + 2t$; the edge labels are unchanged. The smallest feasible value is $t = -214$, which gives $x_1 = 1$ (x_1 is the smallest vertex label), but in this case $x_7 = y_{10} = 100$. Similarly, $t = -213$ forces $x_5 = y_{11} = 133$. The next case, $t = -212$, gives a minimal totally magic injection with $h = 577$, $k = 365$ and deficiency 181. It is shown in Fig. 4.7.

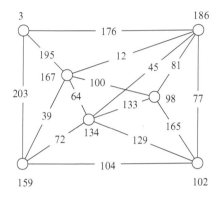

Fig. 4.7. A totally magic injection for G_{1200}

It remains to show that there are no further TMI graphs of order 7. In Table 4.3 we exhibit the vector N_G for the remaining 41 graphs. In each case it is seen that every case N_G contains two equal entries (which implies that two edge labels are equal), a zero (implying a zero edge label) or both positive and negative entries (so that there must be a negative edge label).

So we have:

Theorem 4.22 *There are precisely six TMI graphs with seven vertices, of which two (G_{1200} and $K_{1,6}$) are connected.*

Research Problem 4.5 *How many TMI graphs are there with eight vertices?*

n	N_{G_n}	$-M_{G_n}^-$
839	[0, 3, 3, 0, -6, -3, -6, 3, 6, 6, 3]	3
964	[21, 21, 21, 0, 0, 0, 21, 21, 0, 21, 0, 0]	63
971	[15, 15, 6, 6, 9, 9, 0, 0, 9, 12, 12, 3]	45
980	[7, 21, 12, 1, 4, -6, -2, 9, 27, -10, -1, 8]	39
981	[15, -6, 2, -1, 10, -3, 2, 1, 4, 12, 9, -5]	21
983	[-1, -2, 3, 3, -2, -3, -3, 2, 2, 2, 2, -3]	1
1000	[15, 6, 6, 15, 18, 9, 18, 21, 12, 12, 21, 15]	81
1005	[0, 9, 0, 9, 0, 9, 9, 0, 9, 9, 0, 0]	27
1050	[30, 26, -1, -21-6, 5, -2, 7, 24, -8, 18, 9, -22]	33
1052	[-22, -34, 24, 42, 14, -36, 2, -26, -2, 30, -20, 36, 12]	24
1064	[-28, -22, 30, 18, 8, 14, 24, 18, -24, -12, -2, -20, -8]	6
1090	[-6, -14, -8, 12, -10, -4, 4, -12, -6, -10, 6, -4, -12]	-30
1093	[-20, 10, -15, 10, -20, -35, -30, -30, -35, -25, 5, 5, -25]	-95
1095	[-24, -12, -24, -6, 18, -6, -18, -30-30, -18, 6, 6, 18]	-54
1096	[0, 18, 0, 18, 0, 0, 18, 0, 0, 0, 18, 18, 18]	54
1100	[-21, -7, -7, -7, -7, -21, 21, 0, 0, 21, -14, -14, 7]	-21
1101	[-21, -7, -7, -21, -35, -35, -28, -14, -14, -14, -14,-28,-7]	-119
1104	[-9, -15, -9,-12, -12, -9,-15,-12,-12, -9, -15, -15, -21]	-81
1146	[-70, 15, 15, -70, -90, -5, -90, -30, -80, -80, 30, -45, -55, -55]	-255
1149	[-48, -48, 36, -24, -60, -60,-72, 12, 12, -36, -48, -48, -72, 0]	-216
1150	[-48, -48, -96, -12, -12, -96, -48, -60, -60, -96, -12, -12, -96, -48]	-360
1153	[-36, -21, -78, 21, -18, 45, -48, 36, -60, 3, -36, -3, -66, -33]	-135
1154	[-60, -20, -60, 0, -20, -100, -80, -80, -40, -100, -40, -20, -40, -40]	-340
1155	[-84, -63, -63, -63, -63, -84, 35, 14, -35, -35, 14, -56, -56, -7]	-259
1156	[-63, -14, -14, -63, -70, -21-70, -77, -28, -28, -28, -28, -77, -35]	-301
1159	[-33, -24, -24, -33, -39, -18, -18, -18, -18, -39, -9, -30, -30, -51]	-189
1166	[-76, -32, -60, -28, -32, -60, -28, -48, -12, -20, -16, -68, -100, -64]	-316
1167	[-40, -60, -60, -40, -45, -65, -45, -35, -35, -85, -15, -15, -85, -75]	-345
1168	[-18, -36, -24, -66, -66, -24, -36, -60, -30, -18, -30, -42, -42, -54]	-270
1170	[-49, -49, -49, -49, -49, -49, -49, -49, -49, -49, -49, -49, -49, -49]	-343
1186	[0, -200, 50, -200, 0, -200, -150, -150, -200, 50, -150, 50, -200, -200, 0]	-700
1189	[-200, -152,-162, 30,-24, -128, 78, -154, 14, -90, -208, -144, 24, -144, -80]	-636
1190	[-96, -96, -192, -48, 0, -48, -240, -96, -144, -144, -48, 0, -240, -192, -96]	-816
1194	[-144, -183, -195, -24, -60, -60, -75, -21, -153-21-189, -57, -72, -189, -72]	-741
1199	[-200, -150, -200, -30, -30, -80, -120, -70, -30, -70, -120, -80, -160, -120, -120]	-780
1205	[-175, -100, -100, -175, -25, -100, -25,-225,-225,-25, -100, -225, -100, -100, -175]	-925
1207	[-48, -48, -48, -24, -24, -48, -48, -24, -24, -24, -24, -48, -48, -48, -48]	-288
1209	[-161, -140, -105, -105, -105, -105, -140, -49, -84, -112, -112, -84, -147, -147, -119]	-861
1222	[-400, -240, -180, -320, -320, -180, -240, -160, -100, -160, -20, -420, -20, -480, -480, -80]	-1880
1228	[-336, -240, -384, -180, -180, -384-240, -228, -84, -84, -228, -288, -288, -288, -288, -144]	-1944
1232	[-192, -216, -216, -168, -288, -312, -312, -288, -168, -264, -264, -264, -264, -144, -168, -168]	-1872

Table 4.3. The vector N_G and det M_G^- for 41 survivors G

5

Magic Type Labelings of Digraphs

5.1 Magic Labelings

Early labeling results were all on undirected graphs. However, in the mid-1980s Bloom and Hsu defined graceful labelings on directed graphs [16]. Not much work was done with directed graph labeling until the early 2000s when Marr worked with directed graph labelings in her dissertation [65]. As part of that work, Bloom et al. defined magic labelings on directed graphs [17]. The results in this chapter are from [17] unless otherwise noted.

5.1.1 Definitions

A *magic labeling* of a digraph D is a one-to-one map λ from $V(D) \cup E(D)$ onto the set of integers $\{1, 2, \ldots, v + e\}$ in which all the sums

$$m_A(x) = \lambda(x) + \sum_{x \to y} \lambda(x, y)$$

and all the sums

$$m_B(x) = \lambda(x) + \sum_{z \to x} \lambda(z, x)$$

are constant, independent of the choice of x. A digraph with a magic labeling is a *magic digraph*.

A.M. Marr and W.D. Wallis, *Magic Graphs*, DOI 10.1007/978-0-8176-8391-7_5, 135
© Springer Science+Business Media New York 2013

5.1.2 Basic Counting

Since each label of a magic digraph must be distinct and the magic constant must be the same at every vertex, we immediately recognize that a magic digraph with more than one vertex can have no isolates. In addition, no magic digraph can have a source (or a sink) as this would imply the sum of the edges directed inward to the source (or outward from the sink) would have to equal zero which is not possible since the labels are all positive. Finally, if at any vertex x there was exactly one edge directed inward and one edge directed outward, those two edges would have to have the same label which is not allowed. Hence, we have the following:

Lemma 5.1 *In any magic digraph, each vertex lies on at least three arcs, not all directed into or all directed out of the vertex.*

Theorem 5.2 *In any magic digraph, $e \geq 3v/2$.*

In addition to this edge condition, magic digraphs also have a divisibility condition.

Theorem 5.3 *In any magic digraph, $2v$ must divide $(v + e)(v + e + 1)$.*

Proof. If we add the values $m_A(x)$ (or $m_B(x)$) for all the vertices x, each label will be added exactly once. So,

$$\sum_{x \in V(D)} m_A(x) = \sum_{x \in V(D)} m_B(x) = \sum_{x \in V(D)} \lambda(x) + \sum_{(x,y) \in E(D)} \lambda(x,y)$$

$$= \sum_{i=1}^{v+e} i = \frac{1}{2}(v + e)(v + e + 1). \qquad \square$$

Research Problem 5.1 *Using Theorem 5.3, determine families of graphs that can be shown to not have a magic labeling regardless of the orientation of its edges.*

5.1.3 Small Examples

The one-vertex digraph is trivially magic, and Theorem 5.2 excludes two-vertex cases since we don't allow for loops or repeated arcs.

From Theorem 5.2, a three-vertex magic digraph must have at least five arcs. As there are no repeated arcs, the maximum number of arcs is six. So there are two possible digraphs, and both have magic labelings as shown in Fig. 5.1.

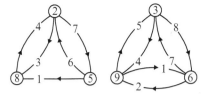

Fig. 5.1. Two smallest magic digraphs

From Theorem 5.2, a four-vertex magic digraph must have at least six arcs; the maximum is 12. Taking Theorem 5.3 into account, there must be 11 or 12 arcs. Again there are two possible digraphs, and again both are magic. Possible labeling arrays are

6	3	12	9
0	5	10	15
11	14	1	4
13	8	7	2

7	4	13	10
1	6	11	16
12	15	2	5
14	9	8	3

For the five vertex case a magic digraph could have 9, 10, 14, 15, 19, or 20 arcs. All cases are possible as shown by the following labeling arrays:

10	4	0	0	7
9	11	1	0	0
0	6	12	3	0
0	0	8	13	0
2	0	0	5	14

11	5	0	0	8
10	12	2	0	0
0	7	13	4	0
0	0	9	14	1
3	0	0	6	15

9	10	19	0	0
0	6	1	14	17
15	0	13	8	2
11	4	0	16	7
3	18	5	0	12

10	11	20	1	8
0	7	2	15	18
16	0	14	9	3
12	5	0	17	8
4	19	6	0	13

5	3	11	19	22
26	6	4	12	15
16	24	7	0	13
14	17	20	8	1
2	10	18	21	9

6	4	12	20	23
24	7	5	13	16
17	25	8	1	14
15	18	21	9	2
3	11	19	22	10

Bloom et al. [13] used computer algorithms to examine the digraphs with 5 vertices and 10 edges. The following nine digraphs were found not to have a magic labeling. No condition has yet been found that explains why these nine specific digraphs could not be magic.

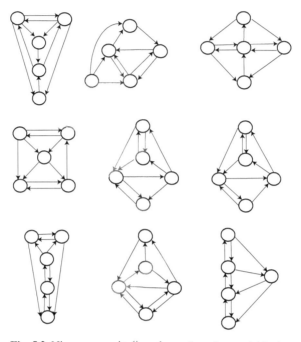

Fig. 5.2. Nine non-magic digraphs on 5 vertices and 10 edges

Research Problem 5.2 *Find reasons why each of the nine digraphs listed in Fig. 5.2 cannot have a magic labeling.*

5.1.4 Complete Digraphs

The complete digraph DK_v has v vertices and all $v(v-1)$ possible arcs. If we consider array for a given DK_v we need each row and column to sum to the same sum and we need to fill each of the v^2 entries with distinct numbers from the set $\{1, 2, \ldots, v^2\}$. Hence, we are only looking for a magic square of order v (we really only need a semi-magic square as the diagonal condition is not necessary). Since these are known to exist for all $v > 2$, we know the DK_v are magic for all $v > 2$.

We can also use the labeling array for DK_v to get a magic labeling for $DK_v - a$ for some arc, a. We begin by permuting the rows and columns so that 1 appears in some entry (i, j) where $i \neq j$. Then certainly the row and column sums will still be equal. If we subtract 1 from each entry, the row and column sums will now all be the same as we have subtracted v from each row and column and the zero entry in position (i, j) in the labeling array will represent the one arc that is to be deleted from DK_v.

5.1.5 Double Cycles

The *double cycle* DC_v is constructed from a v-cycle by replacing each edge by a pair of arcs: edge xy gives rise to arcs (x, y) and (y, x).

Exercise 5.1 *Use Theorem 5.3 to show that DC_v is not magic when v is even.*

Now consider the double cycles where v is odd: say $v = 2k + 1$. Consider the $(2k + 1) \times (2k + 1)$ array A defined by

$$a_{ii} = 4k + i,$$
$$a_{i,i+1} = 2k + 2 - \tfrac{i+1}{2}, i \text{ odd},$$
$$= k + 1 - \tfrac{i}{2}, i \text{ even},$$
$$a_{i+1,i} = 4k + 3 - \tfrac{i+1}{2}, i \text{ odd},$$
$$= 3k + 2 - \tfrac{i}{2}, i \text{ even}$$

(with $a_{1,2k+1} = 3k + 2$ and $a_{2k+1,1} = k + 1$), that is

$$A = \begin{array}{|ccccccc|}
4k+3 & 2k+1 & 0 & 0 & \cdots & 0 & 3k+2 \\
4k+2 & 4k+4 & k & 0 & \cdots & \cdots & 0 \\
0 & 3k+1 & 4k+5 & 2k & 0 & \cdots & 0 \\
0 & 0 & 4k+1 & 4k+6 & k-1 & \cdots & 0 \\
\vdots & \ddots & \ddots & \ddots & \ddots & \ddots & \vdots \\
0 & \cdots & 0 & 0 & 3k+3 & 6k+2 & 1 \\
k+1 & 0 & \cdots & \cdots & 0 & 2k+2 & 6k+3 \\
\end{array}$$

This array contains all the positive integers from 1 to $6k + 3$ once each, and has constant row and column sums $9k + 6$. So it is the labeling array of a DC_{2k+1} with magic constant $9k + 6$. Therefore all double cycles of odd order are magic. The label 1 appears on an arc, so $DC_{2k+1} - a$ is also magic.

An example of the construction is given in Fig. 5.3.

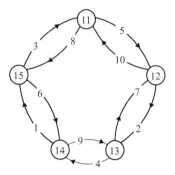

Fig. 5.3. Magic labeling of DC_5

Two other infinite families of regular digraphs that derive from double cycles have also been shown to have a magic labeling. The first are formed from double cycles of odd order by adding a directed spanning cycle. If the double cycle has vertices x_1, x_2, \ldots, x_n, and the directed cycle is $(x_1, x_3, x_5, \ldots, x_n, x_2, x_4, \ldots, x_{n-1})$, a suitable labeling array is

$$
\begin{array}{cccccccc}
4n & 2n+1 & 1 & 0 & \cdots & \cdots & 0 & 2n \\
n+1 & 4n-1 & 3n & 2 & 0 & \cdots & \cdots & 0 \\
0 & n+2 & 4n-2 & 3n-1 & 3 & 0 & \cdots & \\
0 & 0 & n+3 & 4n-3 & 3n-2 & 4 & \cdots & \\
\vdots & \ddots & \ddots & \ddots & \ddots & \ddots & \ddots & \vdots \\
2n+2 & n & 0 & \cdots & \cdots & 0 & 2n-1 & 3n+1
\end{array}
$$

For even n with vertices x_1, x_2, \ldots, x_n, we form a digraph by adding cycles $(x_{n-1}, x_{n-3}, \ldots, x_3, x_1)$ and $(x_n, x_{n-2}, \ldots, x_4, x_2)$. The matrix

$$
\begin{array}{cccccccc}
4n & 2n+1 & 0 & \cdots & \cdots & 0 & n-1 & n+2 \\
n+1 & 4n-1 & 2n+2 & 0 & \cdots & \cdots & 0 & n \\
1 & 2n & 4n-2 & 2n+3 & 0 & \cdots & \cdots & 0 \\
0 & 2 & 2n-1 & 4n-3 & 2n+4 & 0 & \cdots & \\
\vdots & \ddots & \ddots & \ddots & \ddots & \ddots & \ddots & \vdots \\
3n & 0 & \cdots & \cdots & 0 & n-2 & n+3 & 3n+1
\end{array}
$$

provides a labeling. In both cases the subtraction technique may be applied; the resulting magic digraph is formed by deleting one of the new edges.

Examples of these two families are shown in Fig. 5.4.

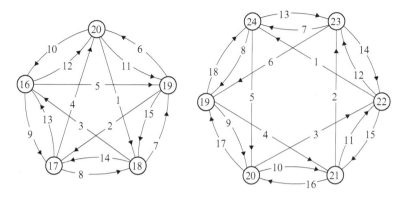

Fig. 5.4. Magic $DC_5 + C_5$ and $DC_6 + 2C_3$

5.1.6 Magic Digraphs with Two Magic Constants

In [65], several other types of magic labelings were mentioned but not explored. One such labeling was a slight modification of magic labelings. Consider the labeling λ from $V(D) \cup E(D)$ into $\{1, 2, \ldots, v + e\}$ with the condition that

$$m_A(x) = [\lambda(x) + \sum_{x \to y} \lambda(x, y)] = k_{\text{in}}$$

and the sums

$$m_B(x) = [\lambda(x) + \sum_{z \to x} \lambda(z, x)] = k_{\text{out}},$$

but k_{in} is not necessarily equal to k_{out}.

Exercise 5.2 *Find an example of such a labeling where $k_{\text{in}} \neq k_{\text{out}}$ or prove such a labeling is not possible.*

5.2 Locally Magic Labelings

5.2.1 Definitions

A *locally magic labeling* or *magic vertex labeling* on a digraph D is a one-to-one map λ from $E(D)$ into a subset of integers $\{1, 2, \ldots, v + e\}$, where $v = |V(D)|$ and $e = |E(D)|$ in which the sum

$$m_A(x) = \sum_{x \to y} \lambda(x, y)$$

and the sum

$$m_B(x) = \sum_{z \to x} \lambda(z, x)$$

are constant for each x, but this constant need not be the same for each vertex x in D. A digraph with a locally magic labeling is called a *locally magic digraph*. A *strong locally magic labeling* of a digraph D is a locally magic labeling in which only the labels $\{1, 2, \ldots, e\}$ are used; digraphs with such a labeling are called *strongly locally magic.*

Note that any digraph that is magic is certainly locally magic. However, the digraph may not be strongly locally magic. This relationship is one of the reasons we include this particular labeling in this text even though it is not a total labeling. It is not true that a digraph that is locally magic is necessarily magic. For example, the digraph in Fig. 5.5 has a strong locally magic labeling, but was shown not to be magic.

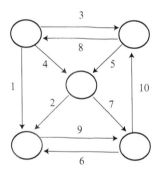

Fig. 5.5. A locally magic digraph

Also, it is useful to note that the edge condition for magic digraphs $(e \geq 3v/2)$ will still hold for locally magic digraphs, but there is no longer a divisibility condition since the sum is not the same across all vertices.

A *conservative labeling* of a graph G is a labeling under which some orientation of G is strongly locally magic. A graph possessing such a labeling is known as a *conservative graph*. Conservative graphs were first defined in 1980 [10] and locally magic labelings were defined in 2007 [66]. However, the connection between these two labelings was not made until 2010 because of the notational differences. Hence, many of the results on conservative graphs parallel those results on locally magic labelings. However, the key difference is that showing a graph is conservative usually only shows there is some orientation that is locally magic, but not necessarily which orientation. Results on locally magic labelings are usually in reference to a specific orientation.

5.2.2 Digraphs with no Locally Magic Labeling

Studying digraphs that are not locally magic is useful, as these digraphs will also be non-magic. This section presents two forbidden structures for all locally magic digraphs.

A digraph D has a *2-cut* if removing any two edges disconnects the underlying graph. Suppose that the underlying graph of a digraph D has an edge cut that produces the connected components G and G'. Then, the edge cut is a *unidirectional cut* or *unicut* if all of the directed edges in the edge cut are from G to G' (or vice versa).

Theorem 5.4 *A digraph with either (a) a unicut or (b) a 2-cut cannot have a locally magic labeling, hence it cannot have a magic labeling either.*

Proof. Suppose a digraph D has a locally magic labeling λ, and suppose D has a unicut disconnecting the digraph into two digraphs D_1 and D_2. Suppose the unicut is directed from D_2 to D_1 and consists of edges e_1, e_2, \ldots, e_n. Examining the vertices in D_1, if we add the amount going in to every vertex in D_1, this should equal the amount going out of every vertex in D_1. However, every edge that is not in the unicut will appear in both of these sums. Thus, we can cancel these labels from both sides of the equation yielding $\sum_{i=1}^{n} \lambda(e_i) = 0$. But, since every edge label is positive, this is impossible. Thus, D cannot have a locally magic labeling.

Suppose D has a 2-cut. Let D_1 and D_2 be the new components. Then, if the 2-cut is a unicut, we are finished by above. Thus, we only need to consider the case where one edge (e_1) is directed into D_1 and one edge (e_2) is directed out of D_1. Then, if we sum all the labels going in to vertices of D_1 and sum all the labels going out of vertices of D_1, these sums should be equal. But, again each edge label except the two in the unicut will appear in both sides of these sums. Hence, we have $\lambda(e_1) = \lambda(e_2)$, which is impossible since λ is injective. Hence, D cannot have a locally magic labeling. $\qquad\square$

Recall that each vertex w of a digraph has an *in-degree* $\mathrm{in}(w)$ (the number of arcs directed inward, toward w) and an *out-degree* $\mathrm{od}(w)$.

Theorem 5.5 *A digraph with a vertex w with $\mathrm{od}(w) = 1$ and $\mathrm{in}(w) = k$ (or vice versa) and $k^2 + k > 2v + 2e$ cannot have a magic vertex labeling.*

Proof. Suppose D has such a vertex. Then, the largest possible label for the edge directed outward is $v + e$. Then, if the edges directed inward are labeled with the smallest possible values, we get that sum to be $(k^2 + k)/2$, but if this sum is greater than $v + e$, we will not be able to produce a locally magic labeling. Hence, the result follows. $\qquad\square$

In particular, a locally magic digraph with a vertex w with $\text{od}(w) = 1$ and $\text{in}(w) = v - 2$ must have $v^2 - 5v + 2 \leq 2e$.

5.2.3 Cycles

Lemma 5.6 *The double cycle DC_n is (strongly) locally magic for all $n \neq 2$.*

Proof. DC_1 is trivially locally magic, while DC_2 is not locally magic by Theorem 5.4. For $n \geq 3$, we simply label each edge on the outer (clockwise) cycle with an odd number. Then, if (x, y) is an edge on the outer cycle with label k, (y, x) gets label $k + 1$. Next, if vertex x is adjacent to w and y with (x, y) labeled k and (w, x) labeled j, we have the indegree and outdegree at x equal to $j + k + 1$. \square

Theorem 5.7 *A digraph that can be decomposed into any collection of double cycles with three or more vertices is (strongly) locally magic.*

Proof. Using Lemma 5.6, we can label any DC_n with $n \geq 3$ with a strong locally magic labeling. Hence, every vertex involved in that cycle will have the locally magic property. If we label each of the cycles in this manner with increasing groups of edge labels, the entire digraph will be strongly locally magic. \square

5.2.4 Wheels

Although Theorem 5.5 shows that many orientations of wheels are not locally magic, there is at least one orientation that is. This was shown in [10] and one specific orientation was given in [66].

A *unicyclic wheel* is a wheel in which the edges in the outer cycle are all oriented in the same direction. A wheel has *alternating spokes* if every other spoke from the center vertex has an opposite orientation.

Exercise 5.3 *Show the unicyclic wheel W_n with alternating spokes has a (strong) locally magic labeling for all $n \geq 3$.*

5.2.5 Complete Graphs

The following theorems will be helpful in proving various classes of graphs are conservative. All results can be found in [10]. A graph with a conservative labeling that produces an orientation in which at each vertex there are an equal number of inwardly and outwardly directed edges is known as an *Eulerian conservative graph*.

Theorem 5.8 *If a graph G can be decomposed into a conservative subgraph H_1 and an Eulerian conservative subgraph H_2, then G is conservative.*

Proof. Suppose H_1 has q_1 edges. A conservative labeling of G can be constructed by using the conservative labeling of H_1 and adding q_1 to each edge label of the conservative labeling of H_2. □

Suppose that a graph G is the sum of two spanning cycles. Then, labeling the edges in one cycle with consecutive odd integers and labeling the edges of the second cycle starting at the same vertex with the even integers in descending order gives the following result.

Theorem 5.9 *If G is the sum of two spanning cycles, then G is Eulerian conservative.*

Exercise 5.4 *Show that if a graph G has $p = 2n + 1$ vertices and is the sum of three spanning cycles a conservative labeling of G can be found.*

Since K_n is known to be a sum of n spanning cycles for n odd, the above results imply that K_n is conservative as long as $n \neq 3$. If n is even (and $n \neq 6$), K_n can be partitioned into a wheel W_{n-1} and K_{n-1}^* where K_{n-1}^* is the complete graph minus one spanning cycle. And since each of these two graphs is conservative, we get that K_n is also conservative.

Exercise 5.5 *Find a conservative labeling for K_6.*

With this exercise and the above comments we get the following theorem.

Theorem 5.10 *The complete graphs K_n are conservative for all $n \geq 4$.*

In addition, the double complete digraph DK_v is locally magic for all $v \geq 3$ since it is magic for all $v \geq 3$.

5.2.6 Complete Bipartite Digraphs

Theorem 5.11 *If G has $2n$ vertices and G can be decomposed into two disjoint spanning cycles and two disjoint cycles of length n whose union spans G, then G is Eulerian conservative.*

Proof. Label the vertices in one of the cycles of length n as v_1, v_2, \ldots, v_n and the remaining vertices as u_1, u_2, \ldots, u_n. Number the edges in the first cycle with the sequence $3, 9, 15, \ldots, 3(2n - 1)$ starting at v_1. Number the edges in the other cycle with the sequence $6, 12, 18, \ldots, 6n$ starting at u_1.

Then number one of the spanning cycles with the integers equal to 1 (mod 3) in descending order beginning at v_1 with $6n - 2$. Number the final cycle with the integers equal to 2 (mod 3) in descending order beginning at u_1 with $6n - 1$. □

Using Theorem 5.9 we find a conservative labeling for $K_{4,4}$ and using Theorem 5.11 we find a conservative labeling for $K_{6,6}$. Figure 5.6 gives a conservative labeling for $K_{4,6}$. Then, applying Theorem 5.8 we get the following result.

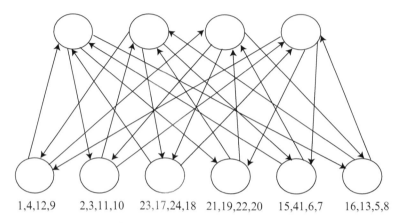

1,4,12,9 2,3,11,10 23,17,24,18 21,19,22,20 15,41,6,7 16,13,5,8

Fig. 5.6. Conservative labeling of $K_{4,6}$

Theorem 5.12 *The complete bipartite graphs $K_{2n,2m}$ are conservative for all $n, m \geq 2$.*

The following theorem gives information on some classes of complete bipartite graphs that cannot be conservative.

Theorem 5.13 *If the edge set of G can be partitioned into disjoint minimal cutsets and $e \equiv 1$ or 2 (mod 4), then G is not conservative*

Proof. Let G be conservatively labeled. Any minimal cutset divides G into two components. Since the amount going in to each component must equal the amount going out of that component, the sum of the edge numbers on this cutset must be even. Since $E(G)$ can be partitioned into disjoint minimal cutsets, the sum of all the edge numbers must be even, but this only occurs when $e \equiv 0$ or 3 (mod 4). □

Since bipartite graphs by definition can be partitioned into disjoint minimal cutsets, we know that $K_{m,n}$ is not conservative when $n \cdot m \equiv 1$ or 2 (mod 4).

Research Problem 5.3 *Which of the remaining complete bipartite graphs $K_{m,n}$ are conservative?*

5.3 Vertex-Magic Edge Labelings

Unless otherwise noted, all the results in this section are from [67].

A digraph D with v vertices and e edges has a *vertex-magic edge labeling* if there is a bijection $\lambda : E(D) \to \{1, \ldots, e\}$ such that for every x:

$$\sum_{(y,x) \in E(D)} \lambda(y, x) = \sum_{(x,y) \in E(D)} \lambda(x, y) = k$$

for some $k \in \mathbb{N}$. If such a labeling exists, the digraph is said to be *vertex-magic*.

This means that the sum of the edge labels going into a vertex is equal to the sum of the edge labels going out of that vertex for every vertex in the digraph. Also, the sum is the same at every vertex in the digraph. This gives us a divisibility condition similar to previous ones.

Theorem 5.14 *If a digraph D with v vertices and e edges is vertex-magic, then* $2v | (e + 1)(e)$.

In addition, there is also an edge condition.

Theorem 5.15 *Every vertex-magic digraph has at least $2v + 1$ edges where v is the number of vertices.*

Proof. Suppose there are two vertices with in-degree 1. Then the two edges directed inward must have the same label, which is impossible. Similarly there cannot be two vertices with out-degree 1. If we consider the minimal number of edges at any vertex to guarantee unique labels, we know there must be at least (without loss of generality) one in and two out. Thus, at all other vertices there would need to be at least two edges going in and two edges going out. But, this would give $2v$ edges directed outward and $2(v - 1) + 1 = 2v - 1$ edges directed inward. Since these two numbers must be the same, we know there would be some vertex with three edges directed inward and two outward. Hence, we would have at least $2v$ edges. Showing that $2v$ edges is not possible is left as an exercise. $\qquad\square$

Exercise 5.6 *Explain why there cannot be a vertex-magic graph on v vertices and $2v$ edges.*

Given the divisibility condition and the edge condition, most small cases are eliminated. The labeling matrices for the two smallest digraphs that satisfy the vertex-magic conditions are shown below.

$v = 5$, $e = 14$, s (sum) $= 21$

$$
\begin{array}{ccccc}
- & 0 & 0 & 8 & 13 \\
6 & - & 10 & 0 & 5 \\
2 & 7 & - & 12 & 0 \\
4 & 14 & 0 & - & 3 \\
9 & 0 & 11 & 1 & -
\end{array}
$$

$v = 5$, $e = 15$, s (sum) $= 24$

$$
\begin{array}{ccccc}
- & 0 & 6 & 13 & 5 \\
3 & - & 14 & 0 & 7 \\
2 & 0 & - & 10 & 12 \\
11 & 9 & 4 & - & 0 \\
8 & 15 & 0 & 1 & -
\end{array}
$$

Only one infinite class of graphs has been shown to be vertex-magic. First, note that DK_n for n even cannot be vertex-magic as if $n = 2k$ then $e = (2k)(2k+1)$ and thus $e(e+1) = (2k)(2k+1)(4k^2 + 2k + 1)$ and by Theorem 5.14, DK_n for n even cannot be vertex-magic.

In [37], sparse semi-magic squares are defined. A *sparse semi-magic square* is an $n \times n$ array containing the entries $1, 2, \ldots, m$ with $m < n^2$ once each and the remainder of its entries 0 with its row and columns having constant sum k. Hence, in order to find a vertex-magic edge labeling for odd DK_n we simply need a sparse semi-magic square with zeros along the main diagonal and the values $1, 2, \ldots, n^2 - n$ in the remaining entries.

Exercise 5.7 *Show that a vertex-magic labeling of DK_3 does not exist.*

Diagonal Kotzig arrays can be used to construct sparse semi-magic squares. A *diagonal Kotzig array* is a $d \times n$ rectangular array $(d \leq n)$ with each row being a permutation of the set $\{1, 2, \ldots, n\}$, all the columns have the same sum and all the forward diagonals have the same sum.

To construct $n \times n$ sparse semi-magic squares used to label DK_n for n odd, we begin with an $(n-1) \times n$ diagonal Kotzig array D (defined below) and append a row of zeros to get the matrix A. Let B be the $n \times n$ matrix with $b_{i,j} = i - 1$ when $i \leq n - 1$ and $b_{i,j} = 0$ for $i > n - 1$. Then $S = A + nB$ will be a matrix with a row of zeros on the bottom. Permuting the rows until the zeros appear along the diagonal creates a sparse semi-magic square that can be used to label DK_n for odd n.

Constructing D depends on if $n - 1 \equiv 0 \pmod 4$ or if $n - 1 \equiv 2 \pmod 4$. If $n - 1 \equiv 0 \pmod 4$, the entries of D are constructed in the following way: Let $n - 1 = 4m$. Then, the first m rows can be any permutation of $1, 2, \ldots, n$. The next m rows are formed by letting $D_{i+m,j} = n + 1 - D_{i,j-1}$ for $1 \leq i \leq m$ (where the subscripts are taken mod n). For the bottom $2m$ rows of D just take the complements of the first $2m$ rows in the same order. More precisely, $D_{i+2m,j} = n + 1 - D_{i,j}$.

If $n - 1 \equiv 2 \pmod 4$ with $n - 1 = 4m + 2$. Then, let the first $m - 1$ rows be any permutation of $1, 2, \ldots, n$. The next $m - 1$ rows are given by $D_{i+m-1,j} = n + 1 - D_{i,j-1}$ for $1 \leq i \leq m - 1$. Let row $2m - 1$ be any permutation of $1, 2, \ldots, n$. Define $D_{2m,j+1} = \frac{n+1}{2} + D_{2m-1,j}$ for $j < \frac{n+1}{2}$ and $D_{2m,j+1} = D_{2m-1,j} - \frac{n-1}{2}$ for $j > \frac{n-1}{2}$ (where the subscripts are taken mod n). Row $2m + 1$ is defined by $D_{2m+1,j+2} = \frac{3}{2}(n+1) - D_{2m,j+1} - D_{2m-1,j}$. Then, the bottom $2m + 1$ rows are the complements of the top $2m + 1$ rows.

An example of constructing a sparse semi-magic square used to label DK_7 is given in Table 5.1.

Theorem 5.16 *The double complete graph DK_n is vertex-magic if and only if n is odd and $n \neq 3$.*

Research Problem 5.4 *Find other infinite families of vertex-magic digraphs.*

5.4 Arc-Magic Labelings

5.4.1 Definitions

Recall that an arc is simply another term used for a directed edge. An *(additive) arc-magic labeling* on a digraph D is a bijective map λ from $V(D) \cup E(D)$ onto the integers $\{1, 2, \ldots, v + e\}$, where $v = |V(D)|$ and $e = |E(D)|$ in which the sum $\lambda(xy) + \lambda(y)$ is constant for every arc xy in D. In [11], all digraphs with an arc-magic labeling are characterized.

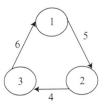

Fig. 5.7. Example of an arc-magic labeling

$$A = \begin{bmatrix} 1\ 2\ 3\ 4\ 5\ 6\ 7 \\ 4\ 5\ 6\ 7\ 1\ 2\ 3 \\ 3\ 1\ 6\ 4\ 2\ 7\ 5 \\ 7\ 6\ 5\ 4\ 3\ 2\ 1 \\ 4\ 3\ 2\ 1\ 7\ 6\ 5 \\ 5\ 7\ 2\ 4\ 6\ 1\ 3 \\ 0\ 0\ 0\ 0\ 0\ 0\ 0 \end{bmatrix} \quad B = \begin{bmatrix} 0\ 0\ 0\ 0\ 0\ 0\ 0 \\ 1\ 1\ 1\ 1\ 1\ 1\ 1 \\ 2\ 2\ 2\ 2\ 2\ 2\ 2 \\ 3\ 3\ 3\ 3\ 3\ 3\ 3 \\ 4\ 4\ 4\ 4\ 4\ 4\ 4 \\ 5\ 5\ 5\ 5\ 5\ 5\ 5 \\ 0\ 0\ 0\ 0\ 0\ 0\ 0 \end{bmatrix}$$

This gives

$$A + 7B = \begin{bmatrix} 1 & 2 & 3 & 4 & 5 & 6 & 7 \\ 11 & 12 & 13 & 14 & 8 & 9 & 10 \\ 17 & 15 & 20 & 18 & 16 & 21 & 19 \\ 28 & 27 & 26 & 25 & 24 & 23 & 22 \\ 32 & 31 & 30 & 29 & 35 & 34 & 33 \\ 40 & 42 & 37 & 39 & 41 & 36 & 38 \\ 0 & 0 & 0 & 0 & 0 & 0 & 0 \end{bmatrix}$$

Permuting the rows to get the zeros along the main diagonal gives the following labeling array for DK_7:

0	36	35	25	20	12	1
7	0	41	29	26	15	11
10	6	0	39	30	27	17
19	9	5	0	37	31	28
22	21	8	4	0	42	32
33	23	16	14	3	0	40
38	34	24	18	13	2	0

Table 5.1. Construction of labeling array for vertex-magic DK_7

The first observation one can make about arc-magic digraphs is that each vertex of D has in-degree at most 1. In order to see this, suppose there is a vertex x with in-degree at least 2 with arcs yx and zx in D. Then, $\lambda(yx) + \lambda(x) = \lambda(zx) + \lambda(x)$, but this would imply $\lambda(yx) = \lambda(zx)$ which is not possible.

Theorem 5.17 *If a digraph D has an arc-magic labeling, then each vertex of D has in-degree at most one.*

5.4.2 Paths and Cycles

Given Theorem 5.17, we now focus our results on paths and cycles as these are the main components of any digraph that would have an arc-magic labeling.

Theorem 5.18 *Suppose that digraph D has an arc-magic labeling. If $v_1, v_2, \ldots,$ v_k is a path in the (undirected) underlying graph G and v_1v_2 is an arc in D, then v_1, v_2, \ldots, v_k is a directed path in D.*

Proof. If v_1, v_2, \ldots, v_k is not a directed path in D, then for some $2 \leq i \leq k-1$ we have that the edges $v_{i-1}v_i$, v_iv_{i+1} are oriented in opposite directions. If both edges are directed into v_i, then v_i has in-degree 2. If both edges are directed out of v_i, then for some $1 \leq j \leq i-1$, we have that v_j has in-degree 2. In either case, we arrive at a contradiction to Theorem 5.17. □

Exercise 5.8 *Suppose digraph D has an arc-magic labeling. If $v_1, v_2, \ldots, v_k, v_1$ is a cycle in the underlying graph G, then show that $v_1, v_2, \ldots, v_k, v_1$ is a directed cycle in D.*

Theorem 5.19 *Suppose digraph D has an arc-magic labeling. Then no component of the underlying graph G contains more than one cycle.*

Proof. Suppose that D has an arc-magic labeling and the underlying graph G has at least two cycles. Then, by Exercise 5.8, these are unidirectional cycles in D. If two cycles share an edge or a vertex, then the component contains a vertex which has in-degree 2. If two cycles in a component are disjoint, then there is a path connecting them. But, by Theorem 5.18, this path is unidirectional and hence D contains a vertex of in-degree 2 where the path joins one of the cycles. Both cases lead to a contradiction of Theorem 5.17. □

5.4.3 Labelings for Arc-Magic Digraphs

This section will describe how to label a given arc-magic digraph and also arrive at the complete characterization of all arc-magic digraphs.

Theorem 5.20 *Every tree T has an orientation that is arc-magic.*

Proof. Let T be a tree with n vertices and hence $n-1$ edges. Select an arbitrary vertex x to serve as the root and orient the tree so that all the edges are directed away from the root. Let $\lambda(x) = 2n-1$. Arbitrarily label each arc with the numbers $1, 2, \ldots, n-1$. Label the remaining vertices v by letting $\lambda(v) = (2n-1) - \lambda(uv)$ where uv is the arc which is directed into v. As each vertex other than x has in-degree 1, this assignment is well defined and all remaining vertices are labeled with the numbers $n, n+1, \ldots, 2n-2$. Thus, we have an arc-magic labeling for this orientation of T. □

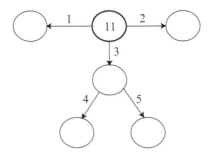

Fig. 5.8. Arc-magic tree labeling in process

Since no vertex can have in-degree larger than 1, the orientations (depending on the choice of the root) described in Theorem 5.20 are the only such orientations that will give an arc-magic labeling.

Exercise 5.9 *Prove that if G is a connected graph with only one cycle, then there is an orientation that will give an arc-magic labeling and find this orientation.*

Theorem 5.21 *If G is a graph such that each component of G has at most one cycle, then G has an orientation that will have an arc-magic labeling.*

Proof. Suppose G has a components. Further, suppose that b components are trees and that $a - b$ components contain exactly one cycle. If G has n vertices, then G has $n - b$ edges.

Orient each acyclic component as outward edges from a root as described in Theorem 5.20. Let v_1, v_2, \ldots, v_b be the roots of these components. Orient each unicyclic component as determined in Exercise 5.9.

Label the roots of the trees with the numbers $2n - b, 2n - b - 1, 2n - b - 2, \ldots, 2n - 2b + 1$. Then, arbitrarily label the arcs of G with the numbers $1, 2, \ldots, n - b$. Label the remaining vertices v in the following way. Let $\lambda(v) = 2n - 2b - 1 - \lambda(uv)$. As the remaining vertices have degree 1, this assignment is well defined and all remaining vertices are labeled with the numbers $n - b + 1, n - b + 2, \ldots, 2n - 2b$. ☐

Again, the orientations generated in all three cases are the only possible orientations with in-degree at most one and thus are all possible orientations that allow an arc-magic labeling.

Putting all these results together gives the following characterization of arc-magic labelings:

Theorem 5.22 *A directed graph D has an arc-magic labeling if and only if every vertex has in-degree either zero or one.*

5.5 In/Out Magic Total Labelings

All the results in this section are from [67] unless otherwise noted.

5.5.1 Basic Ideas

A digraph D with v vertices and e edges is *in-magic* if there exists a bijective function $\lambda : V(D) \cup E(D) \to \{1, 2, \ldots, v + e\}$ such that for every x

$$\lambda(x) + \sum_{(y,x) \in E(D)} \lambda(y, x) = k$$

for some $k \in \mathbb{Z}$. Barone called these labelings (additive) directed vertex-magic labelings [11].

Similarly, a digraph D with v vertices and e edges is *out-magic* if there exists a bijective function $\lambda : V(D) \cup E(D) \to \{1, 2, \ldots, v + e\}$ such that for every x

$$\lambda(x) + \sum_{(x,y) \in E(D)} \lambda(x, y) = m$$

for some $m \in \mathbb{Z}$.

In other words, an in-magic digraph is one where the sum of labels on edges directed in at each vertex is equal, but there is no restriction on the sums going out; and vice versa for an out-magic digraph. Thus any magic digraph is automatically both in-magic and out-magic. But not all graphs that are both in-magic and out-magic are magic.

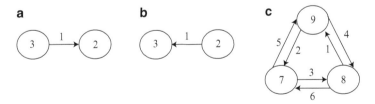

Fig. 5.9. (a) An in-magic digraph, (b) an out-magic digraph, and (c) an in- and out-magic digraph

There is also an interesting relationship between in-magic and out-magic graphs. Given an in-magic digraph, if one reverses the orientations on all the edges the resulting graph will be out-magic and vice versa. The resulting out-magic graph may or may not be isomorphic (without regard to the labeling) to the original in-magic graph. Examples are shown in Fig. 5.10.

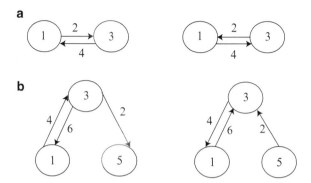

Fig. 5.10. (a) Digraphs that are isomorphic after reversing edges and **(b)** digraphs that are not isomorphic after reversing edges

The following results are given for in-magic graphs, but because of the relationship between in-magic and out-magic digraphs similar results also hold for out-magic digraphs.

Theorem 5.23 *A digraph D with v vertices and e edges is in-magic, if:*

(i) *D satisfies the divisibility condition $2v|(v + e + 1)(v + e)$,*

(ii) *The in-degree sequence of the digraph is a partition of e into v parts such that each part is in $\{0, \ldots, v - 1\}$ and*

(iii) *The smallest number n in the partition satisfies*

$$\sum_{i=0}^{n} v + e - i \geq s$$

where $s = (v + e + 1)(v + e)/2v$.

Proof. (i) Since we are not allowing for repeated edges and loops, the maximum number of edges a vertex p can have going into it is $v - 1$ (one from each vertex). The least amount of edges vertex p can have going into it is 0. Thus, the in-degree sequence of the digraph is a partition of e into v parts such that each part is in $\{0, \ldots, v - 1\}$.

(ii) In order for a digraph to be in-magic, the vertex with the least amount of edges going inward must preserve the magic constant s. Thus, if we take the largest labels possible at that vertex, the sum of those labels must be at least the magic sum. Otherwise, that particular vertex will not have the magic property. □

Theorem 5.24 *A digraph D with v vertices and e edges with only one vertex with in-degree of 0 is said to be in-magic if and only if $v + e = s$.*

Proof. Suppose a digraph D has only one vertex x with in-degree of 0. This means that there are no edges going into x. In order for D to be in-magic, the vertex label of x which is found in the set $\{1, \ldots, v + e\}$ must be equal to s, the sum. Since $v + e$ cannot be greater than s, the only way for D to be in-magic, is if $v + e = s$.

Now, suppose $v + e = s$. Since $v + e$ has to be one of the vertex or edge labels and $v + e$ is the magic constant, this implies that $v + e$ must be a vertex label and one in which no edges are going into that vertex. Hence, that vertex has in-degree of 0. □

Theorem 5.25 *Let $s = (v + e)(v + e + 1)/2v$. Assume s is an integer. Then, a digraph D with more than $(v + e) - k + 1$ vertices of in-degree 1 cannot be in-magic where $k = s/2 + 1$ if s is even and $k = (s + 1)/2$ if s is odd.*

Proof. In order to produce an in-magic labeling of a digraph, the magic sum must be met at each vertex. Let digraph D have m vertices of in-degree 1. Then, we will need m distinct pairs of integers between 1 and $v + e$ that sum to s, the magic constant. However, there is a limit as to the number of such pairs. The smallest two distinct integers that will sum to s are $s/2 + 1$ and $s/2 - 1$ for s even and $(s + 1)/2$ and $(s - 1)/2$ for s odd. Hence, if we count the number of distinct pairs from $v + e$ and $s - (v + e)$ down to these smallest pairs we get the result. □

The previous theorems begin to paint a picture of the conditions necessary for a digraph to be in-magic. However, it would be nice to have both necessary and sufficient conditions for which digraphs are in-magic (and hence out-magic).

5.5.2 Small Digraphs

Table 5.2, below, shows how many digraphs are in-magic, out-magic, or both for all digraphs on less than five vertices. Some cases are not included in the table because of the divisibility condition given in Theorem 5.23. The remaining cases were then examined and digraphs that failed to give unique labels to all the vertices and edges were eliminated. The digraphs that are listed as both in- and out-magic are ones in which when the orientation is reversed, the digraph is isomorphic to itself. The table also shows the number of digraphs which are only in-magic and hence there are an equal number of digraphs that are different from the ones listed as in-magic that are out-magic.

	Number of digraphs	In and out magic	Only in-magic
$v = 1, e = 0$	1	1	0
$v = 2, e = 1$	1	1	0
$v = 2, e = 2$	1	1	0
$v = 3, e = 3$	4	1	1
$v = 3, e = 5$	1	1	0
$v = 3, e = 6$	1	1	0
$v = 4, e = 3$	13	3	4
$v = 4, e = 4$	27	2	4
$v = 4, e = 11$	1	1	0
$v = 4, e = 12$	1	1	0

Table 5.2. Digraphs on less than five vertices

5.5.3 Forbidden Configurations

In [11], in-magic labelings are studied from the perspective of the underlying undirected graph. One benefit to this strategy is that it is easy to find graphs for which no orientation is in-magic. The following results are from [11].

In any orientation of a tree, there is at least one vertex of in-degree 0. If a graph G has at least two components which are trees, then any orientation of G will contain at least two vertices of in-degree 0. But, this would imply both of those vertices would need to have the same label, but that is not possible. This gives the following result.

Theorem 5.26 *Suppose G is a graph which has at least two components which are trees. Then no orientation of G will be in-magic.*

Research Problem 5.5 *Which orientations of trees are in-magic?*

Exercise 5.10 *Use Theorem 5.23 to show that no orientation of wheels, fans, helms, or flower graphs can be in-magic.*

5.5.4 Regular Graphs

In [11], a complete characterization of regular graphs which can be oriented to produce an in-magic labeling is given. Using the divisibility condition, we immediately recognize the relationship between r and n in order for a graph to possibly be in-magic. Then, using constant column sum arrays the remaining cases are eliminated. The result given shows that an orientation exists that makes these graphs in-magic, but does not give all the orientations that do permit an in-magic labeling.

Theorem 5.27 *Suppose that G is an r-regular graph, $r > 0$ with n vertices. Then G can be oriented to permit an in-magic labeling if and only if one of the following four conditions on r and n hold.*

(i) *If $r \equiv 0(\mathrm{mod}\ 4)$ and $n \equiv 1(\mathrm{mod}\ 2)$.*

(ii) *If $r \equiv 1(\mathrm{mod}\ 4)$ and $n \equiv 2(\mathrm{mod}\ 8)$.*

(iii) *If $r \equiv 3(\mathrm{mod}\ 4)$ and $n \equiv 6(\mathrm{mod}\ 8)$.*

(iv) *If $r \equiv 2(\mathrm{mod}\ 4)$.*

Research Problem 5.6 *Find a characterization of all in-magic (out-magic) digraphs.*

Notes on the Research Problems

Edge-Magic Labelings

The first problem,

Research Problem 2.1. *Investigate graphs G for which Equation (2.3) implies the nonexistence of an edge-magic total labeling of $2G$,*

is a simple, open-ended, investigative exercise. As a follow-up, one could look at Sect. 2.2.1 with a view to generalizing it.

The study of magic numbers is almost untouched. Kotzig and Rosa mentioned it in [54], but there are not many results. In **Research Problem 2.2** we ask for $M(7)$ and $M(8)$; these can be found by exhaustive searches. A more general—and much harder—problem is to investigate the function $M(n)$ more generally, and in particular

Research Problem 2.3. *Find good lower and upper bounds on $M(n)$.*

Research Problems 2.4, 2.6–2.13 ask for investigation of the edge-magic properties of various classes of graphs: $K_{2n \setminus n}$, (n, t)-kites with n odd, books $B_{k,n}$, generalized books, double stars, complete multipartite graphs, helms, and flowers. This list could be enlarged to taste. We have included classes that seem to be of interest, because they generalize known cases, because someone has asked (in a paper, a talk, or a seminar) about their magic properties, or because other labelings of these graphs have been investigated. In general, if a class of graphs

A.M. Marr and W.D. Wallis, *Magic Graphs*, DOI 10.1007/978-0-8176-8391-7, 159
© Springer Science+Business Media New York 2013

is being studied, some interesting graph-theoretic properties may be revealed by investigating the magic properties.

If a graph, or family of graphs, has a magic labeling, one can always ask for the spectrum. The following case was specifically included because the problem is addressed in [33]. Some progress has been made on this problem in [69], but full results are not yet known.

Research Problem 2.5. *If v is odd, does C_v have an edge-magic total labeling for every magic sum k satisfying $\frac{1}{2}(5v + 3) \le k \le \frac{1}{2}(7v + 3)$? If v is even, does C_v have an edge-magic total labeling for every magic sum k satisfying $\frac{5}{2}v + 2 \le k \le \frac{7}{2}v + 1$?*

Another spectrum problem that was included (**Research Problem 2.16**) is the conjecture by McQuillan and McQuillan on magic triangles.

Research Problem 2.14 asks whether all trees are edge-magic. This was conjectured in [54], and again in [82]. The problem is interesting partly because it is similar to the conjecture that all trees are graceful, and also because it seems to be difficult (no progress in over 30 years).

We propose three problems about disconnected graphs, **Research Problems 2.15, 2.17** and **2.18**: are $K_2 \cup C_n$, nK_4 (n is even) or tW_n (t odd, and $n \equiv 1 \pmod 4$) edge-magic? Many other questions could be asked. $K_2 \cup G$ is interesting because $K_2 \cup G$ is never *vertex*-magic, $G = C_n$ is a simple case, probably with enough structure to make a general solution possible. We know that an *odd* number of copies of K_4 is not edge-magic, so the even case is worth considering. The wheel case is also a slight variation on the case $n \equiv 3 \pmod 4$ that is known not to be edge-magic.

Many questions could be asked about super edge-magic graphs. **Research Problem 2.19** discusses unions of disjoint cycles in the hope that they will shed some light on super edge-magic disconnected graphs.

Vertex-Magic Labelings

We know that there are graphs that have two vertex-magic total labelings with the same vertex labeling but different edge labelings. No one has yet constructed an infinite family of graphs with this property, so **Research Problem 3.1** asks for one. As a variant, one could require that the graphs be connected.

The next two problems,

Research Problem 3.2. *Prove or disprove that every value of h allowed by (3.18) can be realized as the magic constant of a two vertex-magic total labeling of $K_{m,m}$,*

Research Problem 3.3. *Prove or disprove that every value of h allowed by (3.19) and (3.20) can be realized as the magic constant of a vertex-magic total labeling of $K_{m,m+1}$,*

concern the spectrum of vertex-magic complete bipartite graphs. It is expected that almost all of the possible constants can be realized, with only a few small exceptions. However, this may not be true—maybe there is another constraint waiting to be discovered—and in any event the problems may well be quite difficult.

On the other hand, **Research Problem 3.4** asks for the construction of unions of trees that meet the extreme values in terms of numbers of leaves. Their existence should be relatively easy to prove or disprove.

Finally, we asked

Research Problem 3.5. *Can the method of Theorem 3.23 be applied to some case $G \cup H$ where G and H are nonisomorphic regular graphs?* Who knows?

Totally Magic Labelings

The study of totally magic labelings was slow to develop. Researchers felt that totally magic graphs would be relatively rare, but they did not expect them to be particularly interesting. After a little study, the reaction changed to: "Hey, where are these graphs?" and later to wondering why they are so rare. It seems likely that there should some stronger theorems waiting to be found that would rule out the existence of totally magic labelings for large classes of graphs.

Of course, the real problem concerning totally magic labelings is

Research Problem 4.2. *Are there any further totally magic graphs?*

Totally magic injections are discussed primarily in the hope that they will give some insight into this problem. The two research problems,

Research Problem 4.3. *Find examples in which, when the method of Theorem 4.17 is applied to a minimal totally magic injection of G, the resulting injection of $G \cup K_1$ is not minimal,*

Research Problem 4.4. *What is the total deficiency of $K_{1,s} \cup mK_3$?*

should be relatively easy. The problem

Research Problem 4.5. *How many TMI graphs are there with eight vertices?*

may require a lengthy computer run.

Magic Type Labelings of Digraphs

The first problem,

Research Problem 5.1. *Using Theorem 5.3, determine families of graphs that can be shown to not have a magic labeling regardless of the orientation of its edges,*

is not hard and has many options in terms of classes of graphs to examine.

Research Problem 5.2. *Find reasons why each of the nine digraphs listed in Fig. 5.2 cannot have a magic labeling,*

focuses on the nine digraphs given, solutions to this problem could lead to general forbidden configuration theorems for magic digraphs.

Research Problem 5.3. *Which of the remaining complete bipartite graphs $K_{m,n}$ are conservative?*

has not been solved since it originally appeared in [10] over 30 years ago. However, only a few cases remain, so this problem should be manageable.

The last three problems in this chapter, **Research Problems 5.4–5.6,** all ask for more general results for specific classes of digraphs or to find more infinite families of digraphs with the given property. This is a common question in all labeling problems and there is much to be discovered in these problems since these labelings are new to the scene.

References

[1] J. M. Aldous and R. J. Wilson, *Graphs and Applications: An Introductory Approach*. Springer-Verlag, New York (2000).

[2] W. S. Andrews, *Magic Squares and Cubes*. Dover (1960).

[3] A. Armstrong and D. McQuillan, Vertex-magic total labelings of even complete graphs. *Discrete Math.* **311** (2011), 676–683.

[4] S. Arumugam, G. S. Bloom, M. Miller and J. Ryan, Some open problems on graph labelings, *AKCE Internat. J. Graphs Combin.* **6** (2009), 229–236.

[5] S. Avadayappan, R. Vasuki and P. Jeyanthi, Magic strength of a graph. *Indian J. Pure Appl. Math.* **31** (2000), 873–883.

[6] M. Bača, On magic labelings of grid graphs. *Ars Combin.* **33** (1992), 295–299.

[7] M. Bača, Labelings of two classes of plane graphs. *Acta Math. Appl. Sinica* (English Ser.) **9** (1993), 82–87.

[8] M. Bača and M. Miller, *Super Edge-Antimagic Graphs: A Wealth of Problems and Some Solutions*. Brown Walker, Boca Raton (2007).

[9] M. Balakrishnan and G. Marimuthu, E–super vertex magic labelings of graphs. *Discrete Appl. Math.* **160** (2012), 1766–1774.

A.M. Marr and W.D. Wallis, *Magic Graphs*, DOI 10.1007/978-0-8176-8391-7,
© Springer Science+Business Media New York 2013

[10] D. W. Bange, A. E. Barkauskas, and P. J. Slater, Conservative graphs. *J. Graph Theory* **4** (1980), 81–91.

[11] C. A. Barone. Magic Labelings of Directed Graphs, Masters Thesis, University of Victoria, 2008, 55 pages.

[12] S. S. Block and S. A. Tavares, *Before Sudoku: The World of Magic Squares.* Oxford University Press (2009).

[13] G. S. Bloom, Private communication.

[14] G. S. Bloom and S. W. Golomb, Applications of numbered undirected graphs. *Proc. IEEE* **65** (1977) 562–570.

[15] G. S. Bloom and S. W. Golomb, Numbered complete graphs, unusual rulers, and assorted applications. In *Theory and Applications of Graphs, Lecture Notes in Math.* **642**. Springer-Verlag (1978), 53–65.

[16] G. S. Bloom and D. F. Hsu, On graceful directed graphs. *SIAM J. Alg. Disc. Math.* **6** (1985), 519–536.

[17] G. S. Bloom, A. Marr, and W. D. Wallis, Magic digraphs. *J. Combin. Math. Combin. Comput.* **65** (2008), 205–212.

[18] T. Bohman, A construction for sets of integers with distinct subset sums. *Electronic J. Combinatorics* **5** (1998), #R3.

[19] R. A. Brualdi, *Introductory Combinatorics.* (3rd ed.) Prentice-Hall (1999).

[20] B. Calhoun, K. Ferland, L. Lister, and J. Polhill, Totally magic labelings of graphs. *Austral. J. Combin.* **32** (2005), 47–59.

[21] R. Cattell, Vertex-magic total labelings of complete multipartite graphs. *J. Combin. Math. Combin. Comput.* **55** (2005), 187–197.

[22] D. Craft and E. H. Tesar, On a question by Erdös about edge-magic graphs. *Discrete Math.* **207** (1999), 271–276.

[23] J. Dénes and A. D. Keedwell, *Latin Squares and their Applications.* Academic Press (1974).

[24] T. J. Dickson and D. G. Rogers, Problems in graph theory. V. Magic valuations. *Southeast Asian Bull. Math.* **3** (1979), 40–43.

[25] H. Enomoto, A. S. Llado, T. Nakamigawa and G. Ringel, Super edge-magic graphs. *SUT J. Math.* **34** (1998), 105–109.

[26] P. Erdös, Problems and results from additive number theory. *Colloq. Théorie des Nombres, Bruxelles* (1955), 127–137.

[27] P. Erdös and P. Turán, On a problem of Sidon in additive number theory, and on some related problems. *J. London Math. Soc.* **16** (1941), 212–215; Addendum, **19** (1944), 242.

[28] G. Exoo, A. C. H. Ling, J. P. McSorley, N. C. K. Phillips and W. D. Wallis, Totally magic graphs. *Discrete Math.* **254** (2002), 103–113.

[29] R. M. Figueroa-Centeno, R. Ichishima and F. A. Muntaner-Batle, The place of super edge magic labelings among other classes of labelings. *Discrete Math.* **231** (2001), 153–168.

[30] Y. Fukuchi, Edge-magic labelings of wheel graphs. *Tokyo J. Math.* **24** (2001), 153–167.

[31] J. A. Gallian, A dynamic survey of graph labeling. *Electronic J. Combinatorics* **18** (2011), *Dynamic Survey* #DS6.

[32] F. Göbel and C. Hoede, Magic labelings of graphs. *Ars Combin.* **51** (1999), 3–19.

[33] R. D. Godbold and P. J. Slater, All cycles are edge-magic. *Bull. Inst. Combin. Appl.* **22** (1998), 93–97.

[34] J. Gómez, Solution of the conjecture: If $n \equiv 0 \pmod 4$, $n > 4$, then K_n has a super vertex-magic total labeling. *Discrete Math.* **307** (2007), 2525–2534.

[35] J. Gómez, Two new methods to obtain super vertex-magic total labelings of graphs. *Discrete Math.* **308** (2008), 3361–3372

[36] I. D. Gray, Vertex-magic total labelings of regular graphs, *SIAM J. Discrete Math.* **21** (2007), 170–177.

[37] I. D. Gray, J. A. MacDougall, Sparse Semi-Magic Squares and Vertex Magic Labelings. *Ars Comb.* **80** (2006), 225–242.

[38] I. D. Gray, J. A. MacDougall, J. P. McSorley and W D. Wallis, Vertex-magic labeling of trees and forests. *Discrete Math.* **261** (2003), 285–298.

[39] I. D. Gray, J. A. MacDougall, R. J. Simpson and W. D. Wallis, Vertex-magic total labelings of complete bipartite graphs. *Ars Combin.* **69** (2003), 117–128.

[40] I. D. Gray, J. A. MacDougall and W. D. Wallis, On vertex-magic labeling of complete graphs. *Bull. Inst. Combin. Appl.* **38** (2003), 42–44.

[41] R. K. Guy, *Unsolved Problems in Number Theory.* Second Edition. Springer, New York (1994).

[42] T. R. Hagedorn, Magic rectangles revisited. *Discrete Math.* **207** (1999), 65–72.

[43] T. Harmuth, Ueber magische Quadrate und ähnliche Zahlenfiguren. *Arch. Math. Phys.* **66** (1881), 286–313.

[44] T. Harmuth, Ueber magische Rechtecke mit ungeraden Seitenzahlen. *Arch. Math. Phys.* **66** (1881), 413–447.

[45] N. Hartsfield and G. Ringel, *Pearls in Graph Theory.* Academic Press (1990).

[46] S. M. Hegde and S. Shetty, On magic graphs. *Ars Combin.* **27** (2003), 277–284.

[47] H. Ivančo and Lučkaničová, On edge-magic disconnected graphs. *SUT J. Math.* **38** (2002), 175–184.

[48] P. Jeyanthi and V. Swaminathan, Super vertex-magic labeling. *Indian J. Pure Appl. Math.* **34** (2003), 935–939.

[49] P. Jeyanthi and V. Swaminathan, On super vertex-magic labling. *J. Discrete Math. Sci. Cryptography* **8** (2005), 217–224.

[50] S.-R. Kim and J. Y. Park, On super edge-magic graphs. *Ars Combin.* **81** (2006), 113–127.

[51] A. Kotzig, *On a class of graphs without magic valuations. Reports of the CRM* **CRM–136**, 1971.

[52] A. Kotzig, *On magic valuations of trichromatic graphs. Reports of the CRM* **CRM–148**, December 1971.

[53] A. Kotzig, *On well spread sets of integers. Reports of the CRM* **CRM–161** (1972).

[54] A. Kotzig and A. Rosa, Magic valuations of finite graphs. *Canad. Math. Bull.* **13** (1970), 451–461.

[55] A. Kotzig and A. Rosa, Magic valuations of complete graphs. *Publ. CRM–* **175** (1972).

[56] S.-M. Lee, E. Seah and S.-K. Tan, On edge-magic graphs. *Congressus Num.* **86** (1992), 179–191.

[57] K. W. Lih, On magic and consecutive labelings of plane graphs. *Utilitas Math.* **24** (1983), 165–197.

[58] Y. Lin and M. Miller, Vertex-magic total labelings of complete graphs. *Bull. Inst. Combin. Appl.* **33** (2001),68–76.

[59] S. C. Lopez, F. A. Muntaner-Batle and M. Rius-Font, Bimagic and other generalizations of super edge-magic labelings. *Bull. Austral. Math. Soc.* **84** (2011), 137–152.

[60] J. MacDougall, Vertex-magic labeling of regular graph.*Lecture, DIMACS Connect Institute* (July 18, 2002).

[61] J. MacDougall, M. Miller, Slamin and W. D. Wallis, Vertex-magic total labelings. *Utilitas Math.* **61** (2002), 3–21.

[62] J. A. MacDougall, M. Miller and K. Sugeng, Super vertex-magic lotal labelings of graphs. *Proc. 15th AWOCA* (2004), 222–229.

[63] J. MacDougall, M. Miller and W. D. Wallis, Vertex-magic total labelings of wheels and related graphs. *Utilitas Math.* **62** (2002), 175–183.

[64] J. MacDougall and W. D. Wallis, Strong edge-magic labeling of a cycle with a chord. *Austral. J. Combin.* **28** (2003), 245–255.

[65] A. Marr, Labelings of directed graphs, Ph.D. Thesis, Southern Illinois University at Carbondale, 2007, 63 pages.

[66] A. Marr, Magic vertex labelings of digraphs. *Int. J. Math. Comput. Sci.* **4** (2009), no. 2, 91–96.

[67] A. Marr and S. Stern, Magic labelings of directed graphs, preprint.

[68] B. D. McKay, *nauty User's Guide* (version 1.5). *Technical Report* **TR-CS-87-03.** Department of Computer Science, Australian National University (1990).

[69] D. McQuillan, Edge-magic and vertex-magic total labelings of certain cycles. *Ars Combin.* **91**(2009), 257–266.

[70] D. McQuillan and J. M. McQuillan, Magic labelings of triangles. *Discrete Math.* **309**(2009), 2755–2762.

[71] D. McQuillan and K. Smith, Vertex-magic labeling of odd complete graphs. *Discrete Math.* **305**(2005), 240–249.

[72] J. P. McSorley, Totally magic injections of graphs. *J. Combin. Math. Combin. Comput.* **56** (2006), 65–81.

[73] J. P. McSorley and J. A. Trono, On k-minimum and m-minimum edge-magic injections of graphs. *Discrete Math.* **310** (2001), 56–69.

[74] J. P. McSorley and W. D. Wallis, On the spectra of totally magic labelings. *Bull. Inst. Combin. Appl.* **37** (2003), 58–62.

[75] J. Milhalisin, Vertex magic unions of stars. *Congressus Num.* **177** (2005), 101–107.

[76] M. Miller, J. A. MacDougall, Slamin and W. D. Wallis, Problems in total graph labelings, *Proc. Australasian Workshop on Combinatorial Algorithms*, Perth (1999).

[77] J. Moran, *The Wonders of Magic Squares*. Vintage Books, Random House (1982).

[78] F. A. Muntaner-Batle, Special super edge magic labelings of bipartite graphs. *J. Combin. Math. Combin. Comput.* **39** (2001), 107–120.

[79] J. Y. Park, J. H. Choi and J. H. Bae, On super edge-magic labeling of some graphs. *Bull. Korean Math. Soc.* **45** (2008), 11–21.

[80] N. C. K. Phillips, R. S. Rees and W. D. Wallis, Edge-magic total labelings of wheels. *Bull. Inst. Combin. Appl.* **31** (2001), 21–30.

[81] N. C. K. Phillips and W. D. Wallis, Well-spread sequences. *J. Combin. Math. Combin. Comput.* **31** (1999), 91–96.

[82] G. Ringel and A. S. Llado, Another tree conjecture. *Bull. Inst. Combin. Appl.* **18** (1996), 83–85.

[83] J. Sedláček, Problem 27. *Theory of graphs and its applications.* (Smolenice, 1963), 163–164 Publ. House Czechoslovak Acad. Sci., Prague (1964).

[84] S. Sidon, Satz über trigonometrische Polynome und seine Anwendung in der Theorie der Fourier-Reihen. *Math. Ann.* **106** (1932), 536–539.

[85] Slamin, M. Bača, Y. Lin, M. Miller and R. Simanjuntak, Edge-magic total labelings of wheels, fans and friendship graphs. *Bull. Inst. Combin. Appl.* **35** (2002), 89–98.

[86] B. M. Stewart, Magic graphs. *Canad. J. Math.* **18** (1966), 1031–1059.

[87] A. P. Street and W. D. Wallis, *Combinatorial Theory: An Introduction.* Charles Babbage Research Centre (1977).

[88] K. A. Sugeng and M. Miller, On consecutive edge magic total labeling of graph. *J. Discrete Algorithms* **6** (2008), 59–65.

[89] University of Sydney, Mathematics Enrichment Groups, Exercises for March 26, 1999.

[90] L. Valdés, Edge-magic K_p. Paper delivered at *Thirty-Second South-Eastern International Conference on Combinatorics, Graph Theory and Computing* (Baton Rouge, 2001).

[91] V. G. Vizing, On an estimate of the chromatic class of a p-graph [Russian]. *Discret. Anal.* **3** (1964), 25–30.

[92] W. D. Wallis, *Combinatorial Designs.* Marcel Dekker, New York (1988).

[93] W. D. Wallis *A Beginner's Guide to Graph Theory.* Birkhauser, Boston (2000).

[94] W. D. Wallis, Two results of Kotzig on magic labelings. *Bull. Inst. Combin. Appl.* **36** (2002), 23–28.

[95] W. D. Wallis, Vertex magic labelings of multiple graphs. *Congressus Num.* **152** (2001), 81–83.

[96] W. D. Wallis, E. T. Baskoro, M. Miller and Slamin, Edge-magic total labelings. *Austral. J. Combin.* **22** (2000), 177–190.

[97] W. D. Wallis and R. A. Yates, On Totally Magic Injections. *Austral. J. Combin.* **32** (2005), 339–348.

[98] D. B. West, *An Introduction to Graph Theory.* Prentice-Hall (1996).

[99] K. Wijaya and E. T. Baskoro, Edge-magic total labeling on disconnected graphs, *Proc. Eleventh Australasian Workshop on Combinatorial Algorithms*, University of Newcastle (2000), 139–144.

Answers to Selected Exercises

Exercise 1.1 One example is

$$
\begin{array}{ccc}
2 & 1 & 3 \\
3 & 2 & 1 \\
1 & 3 & 2
\end{array}
\qquad
\begin{array}{ccc}
3 & 1 & 2 \\
1 & 2 & 3 \\
2 & 3 & 1
\end{array}
\qquad
\begin{array}{ccc}
8 & 1 & 6 \\
3 & 5 & 7 \\
4 & 9 & 2.
\end{array}
$$

Exercise 1.2 Say the order is $2n - 1$. To get the Latin squares, circulate

$$(n \ 1 \ 2 \ \ldots \ n-1 \ n+1 \ n+2 \ \ldots \ 2n-1),$$

and back-circulate

$$(1 \ 2 \ \ldots \ n-1 \ n+1 \ n+2 \ \ldots \ 2n-1 \ n).$$

Exercise 1.4 A $t \times t$ magic square and a $t \times m$ Kotzig array will produce a $t \times tm$ magic rectangle. For the first ingredient, t must be at least 3, and for the second, t odd implies m odd. So the construction gives all orders $t \times tm, t \geq 3, t(m-1)$ not odd.

Exercise 2.1 Every vertex of tK_{8m+4} has odd degree. The number of edges is $t(4m+2)(8m+3)$, which is even, and is congruent to $2 \pmod 4$ when t is odd, and the number of vertices is $0 \pmod 4$, so $v + e \equiv 2 \pmod 4$, and Theorem 2.1 applies.

A.M. Marr and W.D. Wallis, *Magic Graphs*, DOI 10.1007/978-0-8176-8391-7,
© Springer Science+Business Media New York 2013

Exercise 2.3 Each d_i equals d, so (2.3) is

$$ke = \sigma_0^{v+e} + \sum (d-1)a_i = \sigma_0^{v+e} + (d-1)\sum a_i = \sigma_0^{v+e} + (d-1)s,$$

which is (2.4). In a regular graph, $e = \frac{1}{2}dv$, and (2.5) follows.

Exercise 2.5 From Theorem 2.14, $v(K_n + tK_1) + e(K_n + tK_1) \geq \rho^*(n)$. But $v(K_n + tK_1) = t + n$ and $e(K_n + tK_1) = \binom{n}{2}$.

Exercise 2.6 From Theorems 2.16 and 2.15, $\rho^*(n) \geq 2\sigma^*(n-1) \geq 2(4 + \binom{n-2}{2})) = 8 + (n-2)(n-3) = n^2 - 5n + 14$.

Exercise 2.7 We use the Fibonacci numbers

$$f_2 = 1, f_3 = 2, \ldots, f_{n+1}.$$

Write $k = f_{n+3} + 1$ and define a labeling λ of the K_n with vertices $\{x_1, x_2, \ldots, x_n\}$ by

$$\lambda(x_i) = f_{i-1}$$

for $i = 1, 2, \ldots, n$ and

$$\lambda(x_i x_j) = k - \lambda(x_i) - \lambda(x_j).$$

The smallest edge label is the label on $x_{n-1}x_n$, namely $f_{n+3} + 1 - f_n - f_{n+1} = f_{n+1} + f_{n+2} + 1 - f_n - f_{n+1} = f_{n+1} + 1$, and this is greater than any vertex label. The largest label used will be at most $k - f_2 - f_3 = k - 3$. So the labels are all positive, and no label is repeated. For each number between 1 and $k - 3$ that has not been used as a label, introduce a new vertex with that number as a label. The result is a magic graph with magic sum k. There will be

$$t = k - 3 - n - \tfrac{1}{2}n(n-1)$$

new vertices, so

$$m(G) \leq f_{n+3} - 2 - n - \tfrac{1}{2}n(n-1).$$

Exercise 2.8 From the computational results on Sidon sequences, we know

$$\rho^*(7) = 30, \quad \rho^*(8) = 43.$$

When $n \geq 8$, the equation

$$\rho^*(n) \geq n^2 - 5n + 14 \text{ when } n > 8$$

(equation (2.18), proven in Exercise 2.6) applies. Suppose an edge-magic total labeling of K_n existed. From Theorem 2.14,

$$n + \binom{n}{2} \geq \rho^*(n).$$

For $n = 7$, this says $28 \geq 30$, and for $n = 8$, it says $36 \geq 43$. Both are false. And provided $n \geq 8$,

$$\tfrac{1}{2}n(n-1) + n \geq n^2 - 5n + 14$$

or equivalently

$$n^2 - 11n + 28 \leq 0,$$

which is false for all such n.

Exercise 2.10 Label the endpoints of the common edge with 1 and $n+2$, and the other vertices $2, 3, \ldots, n+1$.

Exercise 2.11 Take a $K_{1,n+1}$ with center labeled $n + 2$ and other vertex labels $n+1, n+6, n+7, \ldots, 2n+5$. Attach an edge to the $n+1$ and label the other endpoint $n+3$ (or $n+4$). This has $k = 3n+8$. It is by no means unique: a $K_{1,3}$ with center 1 and other labels 5, 8, 9, with 6 attached to the 5, works with $k = 13$.

Exercise 2.12 P is regular of degree 3 with $v = 10, e = 15$. So $2\sigma_0^{10} + \sigma_0^{25} \leq 15k \leq 2\sigma_{15}^{25} + \sigma_0^{25}$, or $435 \leq 15k \leq 735$. So $29 \leq k \leq 49$. Below is an example with $k = 29$, which is in fact super edge-magic.

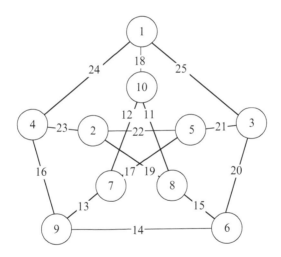

Exercise 2.13 Use the same notation as for wheels; assume that the edge deleted would have received label s_n. Then

$$
\begin{aligned}
c &= n+1 \\
a_i &= \tfrac{1}{2}(i+1) & \text{if } i \text{ is odd,} \\
a_i &= 2n + \tfrac{1}{2}i & \text{if } i \text{ is even,} \\
b_i &= 3n + 1 - \tfrac{1}{2}(i+1) & \text{if } i \text{ is odd,} \\
b_i &= n + 1 - \tfrac{1}{2}i & \text{if } i \text{ is even,} \\
s_i &= 2n - i + 1.
\end{aligned}
$$

is an edge-magic total labeling with $k = 4n + 2$ [85].

Exercise 2.15 Say the one-factor consists of vertices x_i and y_i and edges $x_i y_i$ for $1 \le i \le n$. $\mathrm{wt}(x_i y_i) = \lambda(x_i) + \lambda(y_i) + \lambda(x_i y_i)$. Adding,

$$nk = \sum (\lambda(x_i) + \lambda(y_i) + \lambda(x_i y_i))$$
$$= \sigma_0^{3n} = \tfrac{1}{2} 3n(3n + 1),$$

so $k = \frac{3(3n+1)}{2}$, which is not an integer unless n is odd. For odd n, select a Kotzig array A of size $3 \times n$ and put

$$\lambda(x_i) = a_{1i} + 1, \quad \lambda(y_i) = a_{2i} + n + 1, \quad \lambda(x_i y_i) = a_{3i} + 2n + 1.$$

Exercise 2.16 For a given k, the labeling is essentially determined by the labels on the vertices of the cycle (in order)—there are choices of which label goes on the edge of K_2, but the difference is unimportant. For $K_2 \cup C_3$, (2.2) gives $13 \le k \le 17$. For $k = 13$, the vertex labels sum to 7; the only cycle of labels is $(1, 2, 4)$, which does not work. (A label 11 is required, but 11 is too big.) $k = 14$ gives sum 11, and none of the possible cycles $(1, 2, 8)$, $(1, 3, 7)$, $(1, 4, 6)$, $(2, 3, 6)$, $(2, 4, 5)$ work. If $k = 15$, the sum is 15; the label on a vertex of a triangle would necessarily equal that on the opposite edge. Case $k = 16, 17$ are the duals of $k = 14, 13$. On the other hand, for $K_2 \cup C_4$, one solution with $k = 17$ is given by cycle labels $(1, 4, 3, 10)$.

Exercise 2.17 For example, by the magic property, $x_i + y_i + Z_i = x_i + z_i + Y_i$ which gives $Y_i - y_i = Z_i - z_i$.

Exercise 2.18 There are a few different approaches to this problem. One way is to use modular arithmetic (see [70]).

Exercise 2.20 The solution to Exercise 2.15 gives a super solution.

Exercise 2.21 Suppose λ is a super edge-magic total labeling. Say the edge with label $v + i$ is $x_i y_i$ (a vertex may receive more than one name). Then $\lambda(x_i) + \lambda(y_i) + \lambda(x_i y_i) = k$ so $\lambda(x_i) + \lambda(y_i) = k - v - i$ and

$$\{\lambda(x_1) + \lambda(y_1), \lambda(x_2) + \lambda(y_2), \dots, \lambda(x_e) + \lambda(y_e)\}$$
$$= \{k - v - 1, k - v - 2, \dots, k - v - e\},$$

a set of consecutive integers. Conversely, suppose λ has the given property; say the consecutive integers are $\{t + 1, t + 2, \dots, t + e\}$. Define $\lambda(xy)$ to be $v - t + \lambda(x) + \lambda(y)$ for each edge xy. This extends λ to a super edge-magic total labeling.

Exercise 2.22 Suppose λ is a super-strong edge-magic total labeling for G. If xy is any edge, then the smallest possible value for $\lambda(x) + \lambda(y)$ is $v_1 + 2$ and the largest possibility is $v_1 + (v_1 + v_2)$. From Exercise 2.21 the values taken by $\lambda(x) + \lambda(y)$ when $x \sim y$ is a set of consecutive integers, so the number of these

sums is at most $v_1 + (v_1 + v_2) - (v_1 + 2) + 1 = v_1 + +v_2 - 1 = v - 1$. The sums are obviously all distinct, so there are e of them. So $e \leq v - 1$.

Exercise 2.23 Apply the labels in such a way that 1 and a_2 are on inadjacent vertices.

Exercise 2.24 Let λ be an edge-magic injection of K_v with length $m(v)$, as provided by Theorem 2.47. The restriction of λ to G has the desired deficiency.

Exercise 3.1 Call the labeling λ, and denote its magic constant by h. Suppose x is a vertex of degree 1, y is adjacent to it, and $\lambda(x) = 0$. Then $\lambda(xy) = h$. Then $\mathrm{wt}(y) \geq \lambda(xy) + \lambda(y) > h$, since $\lambda(y)$ must be greater than 0.

Say P_3 has vertex sequence x, y, z. If 0 is a vertex label, $\lambda(y) = 0$, so $\lambda(xy) + \lambda(yz) = \mathrm{wt}(y) = h$. But $\lambda(x) + \lambda(xy) = \mathrm{wt}(x) = h$ also, so $\lambda(x) = \lambda(yz)$. If 0 is an edge label, we may assume $\lambda(yz) = 0$, so $\lambda(xy) + \lambda(y) = \mathrm{wt}(y) = h$, and $\lambda(x) + \lambda(xy) = \mathrm{wt}(x) = h$, so $\lambda(x) = \lambda(y)$.

One labeling of P_4 has vertex labels (in sequence) $6, 5, 3, 4$, and edges (same order) $0, 1, 2$; it has $h = 6$.

Exercise 3.3 The following example has $h = 59$, the maximum possible.

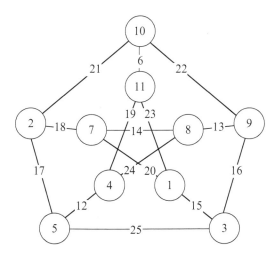

Exercise 3.4 Suppose there is a vertex-magic total labeling with constant h. $3n+3$ labels are used. The sum of the weights of the n vertices of degree 2 is at most the sum of the $3n$ largest labels, so $nh \leq \sigma_3^{3n+3} = \frac{1}{2}(3n + 4)(3n + 3) - 6$. The sum of the weights of the other 2 vertices includes the label of the edge joining them, twice, and $2n + 2$ other labels, so $2h \geq 2 + \sigma_1^{2n+3} = (n + 2)(2n + 3) + 2$. So $n[(n+2)(2n+3)+2] \leq (3n+4)(3n+3) - 12$. It follows that $2n^2 - 2n - 13 \leq 0$. The largest integer solution to this is $n = 3$.

$B_{3,1}$ is C_3 (all cycles are vertex-magic) and $B_{3,2}$ is $K_4 - e$ (a vertex-magic total labeling is given in Fig. 3.1). One VMTL of $B_{3,3}$ is:

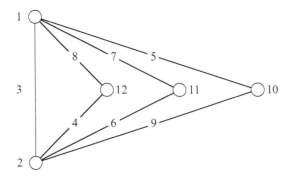

Exercise 3.5 Denote the center vertex of F_n by u and let x_1, \ldots, x_n be the rim vertices. Then

$$
\begin{aligned}
h &= \mathrm{wt}(u) \\
&\geq \sigma_0^{n+1} \\
&= \tfrac{1}{2}(n+1)(n+2).
\end{aligned}
$$

Now consider the sum of weights of all the rim vertices. An upper bound for this sum is found by assigning the $n-1$ largest labels to the rim edges and the $2n$ next largest labels to the rim vertices and the spoke edges. We find

$$
\begin{aligned}
nh &= \mathrm{wt}(x_1) + \cdots + \mathrm{wt}(x_n) \\
&\leq \sigma_1^{2n+1} + 2\sigma_{2n+1}^{3n} \\
&= \sigma_1^{3n} + \sigma_{2n+1}^{3n} \\
&= \tfrac{1}{2}3n(3n+1) - 1 + (2n+1)(n-1) + \tfrac{1}{2}n(n-1) \\
&= \tfrac{1}{2}(9n^2 + 3n - 2 + 4n^2 - 2n - 2 + n^2 - n) \\
&= 7n^2 - 2.
\end{aligned}
$$

$$
h < 7n.
$$

It is easy to see that $7n < \tfrac{1}{2}(n+1)(n+2)$ for all $n > 10$.

Exercise 3.6 Suppose T_n has a vertex-magic total labeling. The weight h of the center vertex is at least $\sigma_0^{2n+1} = (n+1)(2n+1)$. The sum $2nh$ of weights of all the $2n$ rim vertices is at most $2\sigma_{4n+1}^{5n+1} + \sigma_1^{4n+1} = n(17n+9)$. So $2(n+1)(2n+1) \leq 2h \leq 17n+9$, $n \leq \tfrac{1}{8}(11 + \sqrt{233})$. n is integral, so $n \leq 3$. In order to describe the labeling of a triangle in T_n, we write x (a, b, c) y to mean that one spoke receives label x, its endpoint receives a, the rim edge b, other vertex c and other spoke y. Then the following are examples of labelings for $n = 2, 3$:

T_2 : center 3, triangles 8 $(2, 11, 9)$ 1, 4 $(7, 10, 6)$ 5.

T_3 : center 2, triangles 7 $(13, 8, 14)$ 6, 4 $(9, 15, 10)$ 3, 1 $(16, 11, 12)$ 5.

(T_1 is just K_3.)

Exercise 3.7 Examples, with $h = 13$ and 14, respectively, are

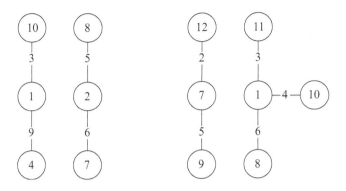

Exercise 3.8 The following is an example with $h = 23$:

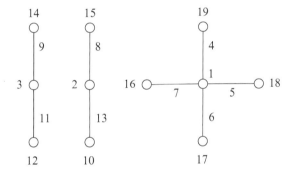

Exercise 3.12 Let the vertices of P_n be v_1, v_2, \ldots, v_n. Then define $\lambda : E(P_n) \cup V(P_n) \to \{1, 2, \ldots, 2n - 1\}$ as follows:

$$\lambda(v_1) = 2n - 1$$
$$\lambda(v_i) = n - 2 + 1 \text{ for } 2 \leq i \leq n$$
$$\lambda(v_i v_{i+1}) = \frac{n - 1}{2} \text{ if } i \text{ is odd}$$
$$= n - \frac{i}{2} \text{ if } i \text{ is even.}$$

Exercise 3.13 Suppose there exist an E-super vertex magic labeling for F_{11}, with magic constant h. Then, $h = 21 + \frac{13}{2} + \frac{(21)(22)}{12} = 66$. If the edges incident

with the center vertex u receives the minimum label 1 to 11, then the sum of the edge label at $u = 66$. Clearly the vertex label at u is greater than or equal to 22 (minimum vertex label). Hence, the sum at vertex u is larger than the magic constant. Thus, F_{11} is not E-super vertex-magic.

Exercise 3.14 The ladder has $v = 2n$ vertices and $e = n + 2(n - 1)$. Then, applying Corollary 3.28.1 $2n$ only divides $2(3n - 2)(3n - 1)$ if $n = 2$. But, $P_2 \times P_2$ is isomorphic to C_4 and by Theorem 3.30 does not have a V-super vertex-magic labeling. Thus, $P_2 \times P_n$ cannot be V-super vertex-magic for all n.

Exercise 4.1 Suppose the center of the star receives label c. Suppose the edges of the star receive labels a_1, a_2, \ldots, a_n, and the vertex adjacent to label a_i receives b_i. Then $a_i + b_i = h = c + \sum a_i$, so b_i is the sum of n positive integers, and therefore $b_i \geq \frac{1}{2}n(n + 1)$ for every i. In particular, If b_1 is the smallest of the b_i and b_n is the largest, then $b_n \geq b_1 + n - 1 \geq \frac{1}{2}n(n+3) - 1$. As $b_n \leq v + e + 2n + 1$, we have $\frac{1}{2}n(n + 3) - 1 \leq v + e + 2n + 1$, so $v + e \geq \frac{1}{2}(n^2 - n - 4)$.

Exercise 4.2 Consider the weights of x and y.

$$h = \lambda(x) + \lambda(xy) + \sum_{\substack{z \sim x \\ z \neq y}} \lambda(xz)$$

$$= \lambda(y) + \lambda(xy) + \sum_{\substack{t \sim y \\ t \neq x}} \lambda(xz).$$

Adding,

$$2h = \lambda(x) + \lambda(xy) + \lambda(y) + \lambda(xy)$$
$$+ \sum_{\substack{z \sim x \\ z \neq y}} \lambda(xz) + \sum_{\substack{t \sim y \\ t \neq x}} \lambda(xz)$$
$$= k + \lambda(xy) + \sum_{\substack{z \sim x \\ z \neq y}} \lambda(xz) + \sum_{\substack{t \sim y \\ t \neq x}} \lambda(xz),$$

giving the result.

Exercise 4.3 From the vertex-magic property,

$$h = \lambda(x_1) + \lambda(xx_1) + \tau_1$$
$$= \lambda(x_2) + \lambda(xx_2) + \tau_2$$
$$\ldots$$
$$= \lambda(x_n) + \lambda(xx_n) + \tau_n,$$

while edge-magic implies

$$\lambda(x_1) + \lambda(xx_1) = \lambda(x_2) + \lambda(xx_2) = \cdots = \lambda(x_n) + \lambda(xx_n) = k - \lambda(x),$$

so

$$\tau_1 = \tau_2 = \cdots = \tau_n = h - k + \lambda(x).$$

Exercise 4.4 Suppose λ is a totally magic labeling of $P_3 \cup nK_3$ with magic sum and constant k and h. Define a labeling μ by $\mu(x) = 1 + \lambda(x)$ for every element x of $P_3 \cup nK_3$. Then add an edge y joining the endpoints of the P_3, and define $\mu(y) = 1$. We claim that μ is a totally magic labeling of $(n + 1)K_3$, with magic sum and constant $k + 3$ and $h + 3$. It is necessary to verify that $wt_\mu(y) = k + 3$. The center vertex c of P_3 must satisfy $\lambda(c) = k - h$. Suppose the other vertices of P_3 receive labels a and b. Then the two edges must be labeled $h - a$ and $h - b$. So $wt_\lambda(c) = (k - h) + (h - a) + (h - b) = k - h - a - b$. Therefore $k - h - a - b = h$, and $a + b = k$. Thus $wt_\mu(y) = (a + 1) + (b + 1) + 1$.

The result of deleting the y and reducing all labels by 1 is the original labeling λ of $P_3 \cup nK_3$.

Exercise 4.5 It follows from Exercise 4.4 that the possible magic constant and sum values are 3 less than the corresponding values for a totally magic labeling of $(n + 1)K_3$. So there is a totally magic labeling of $P_3 \cup nK_3$ with magic constant h and magic sum k if and only if n is even and $h = 9n + \frac{1}{2}(d - 15)$ and $k = 9n + \frac{1}{2}(15 - d)$, where d is a divisor of $3n + 3$.

Exercise 4.6 Suppose there were a totally magic injection λ. Consider the triangle ace. By Theorem 4.9,
$$\lambda(ab) = \lambda(ed) + \lambda(ef).$$
Again, considering triangle bdf,
$$\lambda(ab) = \lambda(cf) + \lambda(ef).$$
So $\lambda(ed) = \lambda(cf)$.

Exercise 4.7 Theorem 4.4 implies that a forest with a totally magic injection must consist of stars and isolates. Since the centers of any two stars would receive the same label, the only possibilities are K_1, $K_{1,n}$ and $K_1 \cup K_{1,n}$. Since $K_{1,1}$ is not allowed, we have the required list. The smallest possible greatest labels are

$$m_t(K_1) = 1, \quad m_t(K_{1,2}) = 5, \quad m_t(K_{1,n}) = \binom{n + 2}{2} - 2, n > 2,$$

$$m_t(K_1 \cup (K_{1,n}) = \binom{n + 2}{2}.$$

Exercise 5.1 Suppose $v = 2k$. Then, $e = 4k$. Hence, $(v + e)(v + e + 1) = (6k)(6k + 1)$ and $2v = 4k$. But, $4k \nmid (6k)(6k + 1)$.

Exercise 5.2 Suppose such a digraph D exists with labeling θ. Then for each $x \in V(D)$, $k_{\text{in}} = \theta(x) + \sum_{(z,x) \in E(D)} \theta(z, x)$ and $k_{\text{out}} = \theta(x) + \sum_{(x,y) \in E(D)} \theta(x, y)$.

We know that the sum of all the vertex labels in D plus the sum of all the edge labels (either coming in or going out) in D are equal. In other words,

$$(\theta(x_1) + \sum_{(x_1,y)\in E(D)} \theta(x_1,y)) + \cdots + (\theta(x_v) + \sum_{(x_v,y)\in E(D)} \theta(x_v,y))$$

$$= (\theta(x_1) + \sum_{(y,x_1)\in E(D)} \theta(y,x_1)) + \cdots + (\theta(x_v) + \sum_{(y,x_v)\in E(D)} \theta(y,x_v)).$$

Since every edge goes into some vertex and goes out of another vertex so we know that every $\theta(x_i, y)$ must be equal to some $\theta(z, x_j)$ $i, j \in \mathbb{Z}$. Thus, every vertex label appears exactly once on each side of the equation.

The equation implies that $\sum_{x\in V(D)} k_{\text{in}} = \sum_{x\in V(D)} k_{\text{out}}$.

Thus, $vk_{\text{out}} = vk_{\text{in}}$

Hence, $k_{\text{out}} = k_{\text{in}}$.

Exercise 5.3 First, suppose n is odd. Let $V(D) = \{v_1, v_2, \ldots v_n, v_{n+1}\}$. Let $c_1 = (v_n, v_1), c_2 = (v_1, v_2), \ldots, c_n = (v_{n-1}, v_n)$ represent the edges of the cycle and $s_1 = (v_{n+1}, v_1), s_2 = (v_2, v_{n+1}), \ldots, s_n = (v_{n+1}, v_n)$ represent the spokes. We begin by labeling the edges of the cycle. Let $(2n - 2, 2n, 1, 2n - 3, 3, 2n - 5, 5, \ldots, n+2, n-2)$ be the labels along the cycle edges c_1 to c_n. Then, the labels for the spokes s_1 to s_n are forced to be $(2, 2n-1, 2n-4, 2n-6, \ldots, 4, n)$.

If n is even, the cycle can be labeled by $(2, 2n, 1, 2n - 3, 3, 2n - 5, 5, 2n - 7, 7, \ldots)$. Then, the spokes are forced to be $(2n - 2, 2n - 1, 2n - 4, 2n - 6, \ldots 4, n - 1)$.

Exercise 5.4 [10] Label the vertices v_0, v_1, \ldots, v_{2n} in the order they appear in the first cycle, and treat the first and second cycles as beginning at v_0, the third at v_n. Number the edges in the first cycle with the multiples of 3 in the sequence $3, 9, 15, \ldots, 3(2n + 1), 6, 12, \ldots, 3(2n)$, with the edge numbered 3 directed outward at v_0. Then number the edges of the second cycle with the integers equal to 1 (mod 3) in descending order, starting at v_0 with $6n + 1$. Finally, number the third cycle with the integers equal to 2 (mod 3) in descending order starting at v_n with $6n + 2$.

Exercise 5.5 One possible labeling is:

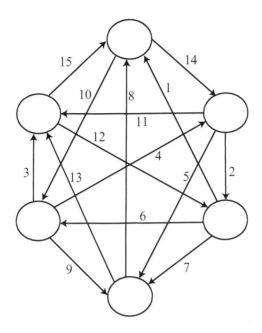

Exercise 5.6 Since $e = 2v$, using Theorem 5.14, we have that the magic constant would be $2v + 1$. However, $2v$ must be a label for some edge and that edge will be involved in both an in-sum and an out-sum. But, there is no way to label the remaining edges so that both in and out sums involving that edge will equal $2v+1$ unless we repeat the label 1 which is not allowed.

Exercise 5.7 This follows immediately from Exercise 5.6.

Exercise 5.8 Suppose that $v_1, v_2, \ldots, v_k, v_1$ is a cycle in the underlying graph G. Without loss of generality, suppose that $v_1 v_2$ is an arc in D. Then we have that v_1, v_2, \ldots, v_k is a path in the underlying graph G, and by Theorem 5.18 we have that v_1, v_2, \ldots, v_k is a directed path in D. By Theorem 5.17, $v_k v_1$ must be an arc in D, otherwise vertex v_k would have in-degree two. The result follows.

Exercise 5.9 Let $C = v_1, v_2, \ldots, v_k, v_1$ be the cycle in G. Consider the graph derived from G by removing the edges $v_1 v_2, v_2 v_3, \ldots, v_k v_1$. What remains is a forest of k trees. Now orient each of the k trees as upbranchings with the vertices v_1, v_2, \ldots, v_k as the roots. To this forest of upbranchings, add the following arcs: $v_1 v_2, v_2 v_3, \ldots, v_k v_1$. Let G' be the name for this orientation of G. By this method, we now have an orientation of G such that the in-degree of every vertex is one. As G' has n vertices and n arcs we use the labels $1, 2, \ldots, 2n$. Arbitrarily

label each arc with the numbers $1, 2, \ldots, 2n$. Label the vertices v in the following way. Let $\lambda(v) = (2n + 1) - \lambda(uv)$ where uv is the arc which has v as its head. As each vertex as in-degree 1, this assignment is well defined and all vertices are labeled with the numbers $n + 1, n + 2, \ldots, 2n$.

Index

A.M. Marr and W.D. Wallis, *Magic Graphs*, DOI 10.1007/978-0-8176-8391-7, 183
© Springer Science+Business Media New York 2013